Geometric Aspects of Probability Theory and Mathematical Statistics

Mathematics and Its Applications

Volume 514

Geometric Aspects of Probability Theory and Mathematical Statistics

by

V.V. Buldygin

Kiev Technical University,
Kiev, Ukraine

and

A.B. Kharazishvili

Tbilisi State University,
Tbilisi, Georgia

SPRINGER-SCIENCE+BUSINESS MEDIA, B.V.

A C.I.P. Catalogue record for this book is available from the Library of Congress.

ISBN 978-90-481-5505-7 ISBN 978-94-017-1687-1 (eBook)
DOI 10.1007/978-94-017-1687-1

The book is a revised and significantly extended version of the title:
V.V. Buldygin, A.B. Kharazishvili "Brunn-Minkowski Inequality and its Applications", Naukova Dumka, Kiev, 1985 (in Russian).
The manuscript was translated from Russian by A.B. Kharazishvili.

Printed on acid-free paper

Table of Contents

Preface

It is well known that contemporary mathematics includes many disciplines. Among them the most important are: set theory, algebra, topology, geometry, functional analysis, probability theory, the theory of differential equations and some others. Furthermore, every mathematical discipline consists of several large sections in which specific problems are investigated and the corresponding technique is developed.

For example, in general topology we have the following extensive chapters: the theory of compact extensions of topological spaces, the theory of continuous mappings, cardinal-valued characteristics of topological spaces, the theory of set-valued (multi-valued) mappings, etc.

Modern algebra is featured by the following domains: linear algebra, group theory, the theory of rings, universal algebras, lattice theory, category theory, and so on.

Concerning modern probability theory, we can easily see that the classification of its domains is much more extensive: measure theory on abstract spaces, Borel and cylindrical measures in infinite-dimensional vector spaces, classical limit theorems, ergodic theory, general stochastic processes, Markov processes, stochastical equations, mathematical statistics, information theory and many others.

On the other hand, there are some fundamental concepts in mathematics which can be met in almost all of its fields and play the main role in various investigations. The standard example of such a concept is the notion of a quasicompact topological space (satisfying the condition that each of its open coverings contains a finite subcovering). This concept turns out to be a necessary tool in contemporary mathematics: many famous theorems essentially utilize it or are closely connected with it. Even in mathematical logic (which seems to be far from topological concepts) this notion is rather important. For example, let us mention that the widely known Gödel theorem on the existence of a model for any logically consistent formal system can be obtained as a consequence of general properties of quasicompact spaces (see, e.g., [172]).

Another important notion which is frequently met in almost all branches of mathematics is the concept of a convex set, and this concept will be carefully considered in our book. Obviously, the convexity is a notion derived from the concept of an algebraic structure (since the convexity is defined in an arbitrary vector space over the field of all real numbers). Nevertheless, some basic properties of convex sets may be regarded as axioms and, in this way, we come to an abstract notion of convexity which is independent of algebraic structures and forms a new type of structure, the so-called structure of abstract convexity.

In our further considerations, we deal only with the standard convexity structure which is canonically induced by the vector space structure. The family of all convex subsets of a given vector space turns out an inner object (i.e., an inner term in the sense of Bourbaki [30]) for the vector structure and, having many remarkable properties, merits to be a subject of extensive study and investigations. The said above can be confirmed by numerous applications of convex sets, not only in mathematics but also in practice. Applications of the theory of convex sets (briefly, convex analysis) are extremely various. Classical Euclidean geometry, classical number theory, mathematical programming, general system theory, mathematical economics, the theory of games, optimization theory and many other fields cannot exist without the developed techniques of the theory of convex sets. Recall, in addition, that basic notions of modern functional analysis are closely connected with the concept of convexity. Furthermore, the most fundamental principles and theorems of this discipline are essentially reliant on convexity (as an illustration, the famous Hahn-Banach theorem on extensions of linear functionals can be pointed out).

However, from the historical viewpoint, the first applications of convex analysis were connected with several inequalities and estimates concerning some combinations of convex functions and convex sets in finite-dimensional Euclidean spaces. Actually, the theory of convex sets in these spaces was founded by Minkowski, who also gave its remarkable applications to certain problems of classical number theory. Moreover, Minkowski introduced new fundamental notions of convex geometry such as: support function of a convex body, support hyperplane, vector sum of convex sets, mixed volumes, and others, which turned out to be very helpful and efficient tools for further investigations. Of course, the above-mentioned notions have played an essential role in establishing important inequalities between various real-valued characteristics of convex bodies. Let us recall that, among such inequalities, the isoperimetric ones constitute an important class. An initial (in some sense) inequality for this class is the so-called Brunn-Minkowski inequality

$$(\lambda_n(A+B))^{1/n} \geq (\lambda_n(A))^{1/n} + (\lambda_n(B))^{1/n},$$

where A and B are arbitrary compact convex bodies in the n-dimensional Euclidean space $\mathbf{R}^n$ and λ_n denotes the standard n-dimensional Lebesgue measure on $\mathbf{R}^n$. It should be indicated that this inequality is reduced to the equality if and only if the given bodies A and B are homothetic (the corresponding homothety coefficient is assumed to be positive).

In our book, we start with the above-mentioned inequality of Brunn and Minkowski. We discuss it, deduce some of its consequences and present applications to probability theory and mathematical statistics. In a certain sense, these questions (and several related topics) form geometric aspects of probability theory and statistics. We believe that they are interesting from the mathematical viewpoint.

The reader must be warned that this book does not contain numerous standard topics concentrated around the Brunn-Minkowski inequality. For instance, various isoperimetric inequalities, Minkowski surface measures, and the properties of different functionals given on the class of convex sets are not considered in detail (these important topics are thoroughly discussed in the remarkable monograph by Hadwiger [85]; see also the books and monographs [47], [88], [178], which were published much later). As pointed out above, our purpose is to demonstrate some less traditional aspects of applications of the convexity inequalities (including the Brunn-Minkowski inequality, of course). Namely, in this book we are oriented to several geometric questions arising in the theory of stochastic processes, to Borel probability measures defined in infinite-dimensional topological vector spaces, and to optimal estimates of random elements. The Table of Contents reflects the main goal of the book and its orientation. We hope that the choice of the material given below is sufficiently justified from the theoretical and practical points of view.

In various respects, the present book is based on the previous one [44] published by the authors in 1985. However, we regard this manuscript as an essentially new book, because many changes were made and extensive additional information is included (new sections, a number of new exercises, presentation of relatively recent results, comments and discussion, extension of Bibliography). Also, the existence of this additional material forced us to change the title from the previous book, since that title seemed to be rather narrow in scope.

In this book, we utilize fairly standard facts and results from mathematical analysis, general topology, convex geometry, measure theory and probability theory. Some of these facts and results are applied without any explanation. Therefore we refer the reader to numerous textbooks and monographs where the corresponding preliminary notions and results from those mathematical disciplines are introduced and discussed sufficiently thoroughly (see, e.g., [29], [31], [57], [59], [74], [80], [85], [86], [107], [124], [131], [154], [155], [171], [173], [175], [180]). Nevertheless, we maximally tried to make our book self-contained and, motivated by this aim, included the material describing several auxiliary facts from the above-mentioned areas of mathematics. This circumstance enables us to assert that the book (being, in fact, a research monograph) is also oriented to graduate and postgraduate students, and to a wide audience of those mathematicians who are interested in deep interactions between different branches of mathematics. In particular, we hope that our small book vividly demonstrates such interactions between the following three fundamental concepts: quasicompactness, convexity and measurability.

Concluding this preface, we wish to stress once more that there are many common topics in classical and modern mathematics. For instance, it will be demonstrated below that some rather elementary inequalities, connected with various combinations of convex bodies in a finite-dimensional Euclidean space, imply interesting and important statements for probability distributions in infinite-dimensional topological vector spaces. This fact shows, in particular, that deep relationships between classical mathematics and modern mathematical theories lead to new nontrivial results and, hence, remain fruitful for further studies.

V.V. Buldygin,
A.B. Kharazishvili

1. Convex sets in vector spaces

This preliminary section of our book is devoted to some general facts and theorems concerning convex sets in vector spaces. They will be helpful in our further considerations. Most theorems presented in the section will be given without proofs. For more information around this topic, we refer the reader to the fundamental monograph by Bourbaki [29] (see also [179]).

The vector spaces considered below are assumed to be over the field $\mathbf{R}$ of all real numbers. It is natural because we are interested in various useful properties of convex sets lying in such spaces. In other words, we will investigate the convex structure deduced from the vector one.

First of all, we wish to recall the definition of a convex set in a vector space. We have already mentioned in the Preface that this notion plays an extremely important role in many domains of mathematics.

If E is a vector space and x and y are any two points of E, then the set

$$[x, y] = \{tx + (1-t)y \ : \ 0 \leq t \leq 1\}$$

is usually called the (linear) segment with the endpoints x and y.

Definition 1. Let E be a vector space and let X be a subset of E. We say that X is convex if, for any two points x and y from X, we have $[x, y] \subset X$.

The following sets provide simple examples of convex subsets of E:

a) empty set ($= \emptyset$) and the whole space E;

b) a one-element subset of E;

c) each segment in E;

d) any linear affine manifold in E (in particular, any affine hyperplane in E);

e) each half-space of E determined by an affine hyperplane in E;

The next two properties of convex sets follow directly from the definition.

1. Let $(X_i)_{i \in I}$ be an arbitrary family of convex sets in E. Then their intersection $\cap_{i \in I} X_i$ is a convex set, too.

2. Let $(X_i)_{i \in I}$ be a family of convex sets in E such that, for any two indices $i \in I$ and $j \in I$, there exists an index $k \in I$ satisfying the relation

$$X_i \cup X_j \subset X_k.$$

Then the union $\cup_{i \in I} X_i$ is a convex set, too.

Definition 2. Let E be a vector space and let X be a subset of E. The convex hull of X (denoted by $conv(X)$) is defined by the formula

$$conv(X) = \cap\{Y \ : \ X \subset Y \subset E, \ Y \ is \ convex \ in \ E\}.$$

In other words, $conv(X)$ is the smallest (with respect to inclusion) convex set containing X.

Example 1. Let $\mathbf{R}^n$ denote the n-dimensional Euclidean space. A set $P \subset \mathbf{R}^n$ is called a convex n-dimensional polyhedron if P is the convex hull of a finite family of points which do not lie in an affine hyperplane of $\mathbf{R}^n$. Another equivalent definition can be formulated as follows. A subset P of $\mathbf{R}^n$ is a convex n-dimensional polyhedron if P is bounded, has nonempty interior and can be represented as the intersection of a finite family of closed half-spaces in $\mathbf{R}^n$. It can easily be observed that the first definition of a polyhedron is formulated only in terms of affine geometry. The second definition appeals also to the topological structure of $\mathbf{R}^n$. Nevertheless, as mentioned above, these two definitions are logically equivalent and the reader can verify this equivalence as a useful exercise.

If we deal with an arbitrary (possibly, infinite-dimensional) vector space E, then it is more convenient to utilize the first definition of a convex polyhedron, i.e. to define a convex polyhedron P in E as the convex hull of a nonempty finite family of points lying in E. In this case, we say that P is n-dimensional if n coincides with the dimension of the smallest (with respect to inclusion) affine linear manifold in E containing P. Sometimes, the empty set $\emptyset$ is also regarded as a polyhedron whose dimension is -1. Further, in the usual way, the notion of a face of a convex polyhedron P can be introduced. We leave to the reader the precise formulation of this notion.

An important particular case of a convex polyhedron is a simplex defined as follows.

Let E be a vector space and let $\{x_0, x_1, ..., x_k\}$ be a finite subset of E whose cardinality equals $k + 1$. Suppose, in addition, that the dimension of the smallest affine linear manifold containing $\{x_0, x_1, ..., x_k\}$ is equal to k (this obviously means that the vectors $x_1 - x_0$, $x_2 - x_0$, ..., $x_k - x_0$ are linearly independent). Then the set

$$S = conv\{x_0, x_1, ..., x_k\}$$

is called a simplex. The points $x_0, x_1, ..., x_k$ are vertices of this simplex and k is its dimension. The empty set is regarded as a simplex, too, and, as mentioned above, its dimension is -1.

In the obvious way, we can define the faces of a given simplex S. Namely, for each subset X of $\{x_0, x_1, ..., x_k\}$, the set $conv(X)$ is a face of our simplex. In particular, the empty set is a face of S and the whole simplex S is a face of itself, too. One-dimensional faces of S, i.e. the segments of the form $[x_i, x_j]$, where $i \neq j$, are called edges of our simplex.

It is convenient to introduce the standard notation for the k-dimensional simplex $S = conv\{x_0, x_1, ..., x_k\}$. In our book we prefer the notation

$$S = [x_0, x_1, ..., x_k]$$

because it is consistent with the notation of a linear segment.

Example 2. Let M be a subset of the Euclidean space $\mathbf{R}^n$. We say that M is a polyhedron in $\mathbf{R}^n$ (not necessarily convex) if it can be represented in the form

$$M = \cup_{i \in I} M_i$$

where $(M_i)_{i \in I}$ is a finite family of convex polyhedra lying in the same space.

It is not hard to show that a set T is a polyhedron in $\mathbf{R}^n$ if and only if it admits a representation in the form

$$T = \cup_{j \in J} T_j$$

where $(T_j)_{j \in J}$ is a finite family of simplices in $\mathbf{R}^n$. We leave an easy proof of this fact to the reader.

It is reasonable to point out once more that the concept of a convex set in a vector space is purely algebraic and, at first sight, does not depend on topological concepts. Nevertheless, for development of convex analysis, topological methods and constructions play very significant role. Moreover, the interplay between the vector and topological structures turns out to be rather helpful for obtaining many nontrivial results and facts in the theory of convex sets.

It is clear that if we want to apply topological methods to convex sets, we are forced to assume some consistency conditions for vector and topological structures (in other words, these structures have to be compatible). This can be reached with the aid of the notion of a topological vector space. We wish to recall here the corresponding definition (see, e.g., [29] or [179]).

Definition 3. Let E be a set equipped with a vector structure and a topological structure. We say that E is a topological vector space if the vector operations

$$(x, y) \to x + y \qquad (x \in E,\ y \in E),$$

$$(t, x) \to tx \qquad (t \in \mathbf{R},\ x \in E)$$

are continuous.

Let us emphasize that, as a rule, topological vector spaces under consideration below are assumed to be Hausdorff.

Example 3. Let E be an arbitrary finite-dimensional (Hausdorff) topological vector space and let $n = dim(E)$ denote its dimension. Then it can be demonstrated that E is isomorphic to the canonical space $\mathbf{R}^n$. In other words, the family of all Euclidean spaces $\mathbf{R}^n$ $(n \in \mathbf{N})$ realizes the class of all finite-dimensional (Hausdorff) topological vector spaces. The reader can try to establish this important fact by induction on $dim(E)$. A stronger result (for vector spaces over more general fields) can be found in [29].

For infinite-dimensional vector spaces, the situation is essentially different, i.e. if E is an infinite-dimensional vector space, then there are many Hausdorff topologies compatible with the vector structure of E (see again [29] or exercises to this section).

The following geometric notions are extremely important in the theory of topological vector spaces.

Definition 4. Let E be a topological vector space, let X and Y be some subsets of E and let H be a closed affine hyperplane in E. We say that H separates X and Y if these sets lie in distinct closed half-spaces determined by H. Furthermore, we say that H strongly separates X and Y if these sets lie in distinct open half-spaces determined by H.

The following statement can be regarded as a geometric form of the famous Hahn-Banach theorem on extensions of linear functionals (see, e.g., [29], [59], [124], [175]).

Theorem 1. *Let E be a topological vector space and let A and B be any two convex subsets of E satisfying the relations:*

1) $A \cap B = \emptyset$;

2) at least one of these sets is open.

Then there exists a closed affine hyperplane in E separating A and B. In addition, if both these sets are open, then there exists a closed affine hyperplane in E strongly separating them.

As an immediate consequence of this theorem, we get the following statement: if A is an open convex subset of E and x is a boundary point for A, then there exists an affine hyperplane H in E passing through x and having no common points with A (evidently, H is closed in E). Actually, H turns out to be a support hyperplane for the closure of A, passing through x (see Definition 6 and Theorem 4 below).

For the sake of completeness, we also present the standard analytic form of the Hahn-Banach theorem.

Theorem 2. *Let E be a vector space and let*

$$p \,:\, E \to \mathbf{R}$$

be a functional on E satisfying the conditions:

$$p(x+y) \le p(x) + p(y) \qquad (x \in E,\ y \in E),$$

$$p(tx) = tp(x) \qquad (x \in E,\ t \ge 0).$$

Suppose, in addition, that a linear functional

$$g_0 \,:\, F \to \mathbf{R}$$

is given on some vector subspace F of E, such that

$$g_0(x) \le p(x) \qquad (x \in F).$$

Then there exists at least one linear functional

$$g \,:\, E \to \mathbf{R}$$

extending g_0 and satisfying the relation

$$g(x) \le p(x) \qquad (x \in E).$$

Theorem 2 is usually applied in those cases when p is some seminorm given on a vector space E (cf. [59] and [124]). In this connection, let us recall that a seminorm on E is any mapping

$$p \,:\, E \to [0, +\infty[$$

satisfying the conditions

$$p(x+y) \leq p(x) + p(y) \qquad (x \in E,\ y \in E),$$

$$p(tx) = |t|p(x) \qquad (t \in \mathbf{R},\ x \in E).$$

If, in addition, the relation $p(x) = 0$ is equivalent to $x = 0$, then p is called a norm on E.

An important class of topological vector spaces is described by the following definition.

Definition 5. We say that a topological vector space E is locally convex if each point of E has a fundamental system (i.e. local base) of convex neighbourhoods.

Evidently, in this definition it is sufficient to require only the existence of a local base of convex neighbourhoods for the neutral element of E.

Another characterization of locally convex topological vector spaces looks as follows: a topological vector space E is locally convex if and only if its topology can be defined by a family of seminorms (see, e.g., [59]). If the topology of E is given by a norm, then E is called a normed vector space. Finally, according to the classical definition, any complete normed vector space is called a Banach space.

Notice that the importance of locally convex topological vector spaces is implied by the fact that there are sufficiently many linear continuous functionals on such spaces. To be more precise, consider any topological vector space E and denote by E^* the family of all continuous linear functionals on E. Obviously, E^* forms a vector space with respect to the natural algebraic operations. This vector space is usually called the conjugate (adjoint or, sometimes, dual) space for E. Let G be a subset of E^*. We say that G separates the points of E if, for any two distinct points $x \in E$ and $y \in E$, there exists at least one functional $g \in G$ such that $g(x) \neq g(y)$.

The next simple (but useful) statement is true.

Theorem 3. *Let E be a locally convex topological vector space. Then E^* separates the points of E.*

This result is one of the immediate consequences of the geometric form of the Hahn-Banach theorem. In connection with Theorem 3, let us note that the converse assertion does not hold in general. In other words, there are examples of a topological vector space E such that E^* separates the points of E, but E is not locally convex.

Definition 6. Let E be a vector space and let X be a subset of E. We say that an affine hyperplane H in E is a support hyperplane for X if

1) $X \cap H \neq \emptyset$;
2) all points of X lie in one closed half-space determined by H.

Definition 7. Let E be a topological vector space and let X be a subset of E. We say that X is a convex body (in E) if X is closed, convex and has nonempty interior.

Example 4. Let $\mathbf{R}^n$ denote, as usual, the n-dimensional Euclidean space and let X be a closed convex subset of this space. Then X is a convex body in $\mathbf{R}^n$ if and only if the smallest affine linear manifold containing X coincides with $\mathbf{R}^n$. In particular, any n-dimensional simplex is a convex body in $\mathbf{R}^n$ (more generally, any n-dimensional convex polyhedron in $\mathbf{R}^n$ is a convex body in $\mathbf{R}^n$).

For convex bodies in topological vector spaces, we have the following two important statements which will frequently be applied (without references) in our further considerations.

Theorem 4. *Let E be a topological vector space and let X be a convex body in E. Then:*
1) each support hyperplane for X is closed in E;
2) for any boundary point $x \in X$, there exists at least one support hyperplane for X passing through x.

Theorem 5. *Let E be a topological vector space and let X be a convex body in E. Then we have the equality*

$$X = \cap_{i \in I} P_i,$$

where $(P_i)_{i \in I}$ denotes the family of all those closed half-spaces in E which are determined by support hyperplanes for X and each of which contains X.

Actually, the two results presented above can easily be deduced from the geometric form of the Hahn-Banach theorem.

Example 5. Evidently, there are many compact convex bodies in the Euclidean space $\mathbf{R}^n$ (for instance, all n-dimensional convex polyhedra and all closed balls in this space are such bodies). Further, a body X in $\mathbf{R}^n$ is compact if and only if X is bounded. On the other hand, let E be a topological vector space in which there exists at least one compact body. Then it can easily be observed that E is locally compact and, consequently, E is finite-dimensional, i.e. isomorphic to $\mathbf{R}^n$ for some natural number n. We thus see that there are no compact bodies in infinite-dimensional topological vector spaces.

In connection with various properties of support hyperplanes for compact sets, let us point out one simple but rather useful statement.

Theorem 6. *Let E be a topological vector space and let X be a nonempty compact subset of E. Then, for each closed affine hyperplane Γ in E, there exists at least one affine hyperplane Γ_1 parallel to Γ and support for X.*

Proof. The argument is very easy. Since our Γ is closed, it is determined by a linear continuous functional

$$g : E \rightarrow \mathbf{R},$$

i.e. for some $r \in \mathbf{R}$, the hyperplane Γ is defined by the formula

$$\Gamma = \{x \in E : g(x) = r\}.$$

Now, let us put

$$r_1 = sup_{x \in X} g(x), \quad r_2 = inf_{x \in X} g(x).$$

Since g is continuous on the compact set X, the values r_1 and r_2 belong to $g(X)$. Consequently, the hyperplanes Γ_1 and Γ_2 defined by the equations

$$g(x) = r_1, \quad g(x) = r_2$$

are support for X and, clearly, they are parallel to the original hyperplane Γ. Of course, the case when the latter two hyperplanes coincide cannot be omitted. This case is precisely that one when our set X lies in some affine hyperplane parallel to Γ.

Theorem 6 implies, in particular, that if X is a compact convex body in a finite-dimensional topological vector space E, then there exist exactly two support hyperplanes for X each of which is parallel to a given hyperplane Γ in E.

Let T be a compact convex body in $\mathbf{R}^n$, where $n \geq 1$, and suppose that $0 \in int(T)$. Fix a unit vector e in $\mathbf{R}^n$ and consider the half-line $l(e)$ in the same space, whose direction coincides with e and whose end-point is 0. Then, according to Theorem 6, there exists a unique hyperplane $\Gamma(e)$ in $\mathbf{R}^n$ such that:

a) $\Gamma(e)$ is a support hyperplane for T;

b) $\Gamma(e)$ is orthogonal to e;

c) $\Gamma(e)$ intersects $l(e)$.

Denote by $h(e)$ the distance between 0 and $\Gamma(e)$. Clearly, in this manner we obtain a certain function $h = h_T$ acting from the unit sphere of $\mathbf{R}^n$ into the open interval $]0, +\infty[$. The above-mentioned function is usually called the Minkowski support function for T. Note that h completely determines the body T. In other words, if T_1 and T_2 are any two compact convex bodies in $\mathbf{R}^n$ such that

$$0 \in int(T_1), \qquad 0 \in int(T_2),$$

then the following assertions are equivalent:

1) the Minkowski support function for T_1 is identical with the Minkowski support function for T_2;

2) $T_1 = T_2$.

The reader can easily deduce the equivalence of 1) and 2) from the separation principle for convex sets (here it is sufficient to apply the weak version of this principle, stating that if a point does not belong to a closed convex set, then there exists a hyperplane strongly separating the point and the set).

The next notion plays the key role in modern convex analysis.

Definition 8. Let E be a vector space and let Z be a convex subset of E. We say that a point $z \in Z$ is extreme for Z if there exists no nondegenerate linear segment in Z whose midpoint is z.

The set of all extreme points for Z will be denoted by the symbol $extr(Z)$.

Example 6. Let $Z = [x_0, x_1, ..., x_k]$ be any k-dimensional simplex in a vector space E. Then it is not hard to see that

$$extr(Z) = \{x_0, x_1, ..., x_k\}.$$

An analogous result holds true for any convex polyhedron P in the space E. Actually, quite often the vertices of P are defined as its extreme points. Such a definition is completely compatible with our geometric intuition.

Example 7. Let B denote the closed unit ball in the space $\mathbf{R}^n$, where $n \geq 1$, and let S be the boundary of B. Then it is evident that $extr(B) = S$. If now E is a finite-dimensional normed vector space and B denotes the closed unit ball in E, then there always are extreme points of B. It is interesting to note that this fact fails to be true for infinite-dimensional Banach spaces. Namely, it may happen that there is no extreme point of the closed unit ball in such a space (the verification of this fact is left to the reader as an exercise).

Definition 9. Let E be a topological vector space and let X be a subset of E. The closed convex hull of X is the intersection of all closed convex sets in E containing X (i.e., is the smallest closed convex set containing X).

Note that the closure of any convex subset of E is convex, too. By starting with this simple observation, it can easily be checked that the closed convex hull of a set $X \subset E$ coincides with $cl(conv(X))$.

One of the most important results in convex analysis is the following theorem due to Krein and Milman (see [126]).

Theorem 7. *Let E be a locally convex topological vector space and let X be a compact convex subset of E. Then X coincides with the closed convex hull of the set of all extreme points of X, i.e.*

$$X = cl(conv(extr(X))).$$

This result implies, in particular, that if X is an arbitrary nonempty compact convex subset of E, then there exists at least one extreme point of X. Moreover, in each support hyperplane of X there is at least one extreme point of X. Let us stress that the Krein-Milman theorem is valid in locally convex topological vector spaces and is not true, in general, for arbitrary topological vector spaces.

On the other hand, for finite-dimensional topological vector spaces, this theorem can be essentially strengthened. Namely, we have the following classical result established by Minkowski.

Theorem 8. *Let E be a finite-dimensional topological vector space and let X be a compact convex subset of E. Then X coincides with the convex hull of the set of all extreme points of X, i.e.*

$$X = conv(extr(X)).$$

The proof of Theorem 8 can be carried out by induction on $dim(E)$ and is not difficult. We leave the corresponding details to the reader. Note also that, for the proof of the Krein-Milman theorem, essentially nonelementary methods are necessary. As a rule, they are based on the principle of transfinite induction or on the Zorn lemma (see, e.g., [29] and [179]).

The next statement is relative to the Krein-Milman theorem and turns out to be rather helpful in certain situations.

Theorem 9. *Let E be a locally convex topological vector space and let Y be a compact subset of E. Suppose, in addition, that the closed convex hull of Y is also compact in E. Then each extreme point of $cl(conv(Y))$ belongs to Y, i.e.*

$$extr(cl(conv(Y))) \subset Y.$$

For the proof, see e.g. [29].

Example 8. Let T be an arbitrary closed convex set in the plane $\mathbf{R}^2$. It is easy to show that the set of all extreme points of T is closed in $\mathbf{R}^2$, too. Consequently, if T is compact, then the set $extr(T)$ is also compact. This geometric fact fails to be true for the space $\mathbf{R}^n$ where $n > 2$. In other words, there exists a compact convex body T in $\mathbf{R}^n$, where $n > 2$, for which the set $extr(T)$ is not closed in $\mathbf{R}^n$. We suggest the reader to construct such a body in the space $\mathbf{R}^3$.

Let us return, for a while, to arbitrary vector spaces (over $\mathbf{R}$) and to convex sets lying in them.

Let E be a vector space and let X be a subset of E. In many cases, we need a more descriptive characterization of the convex hull of X. Some valuable information on the structure of $conv(X)$ yields the equality

$$conv(X) = \cup_{i \in I} conv(X_i)$$

where $(X_i)_{i \in I}$ denotes the family of all finite subsets of X. The validity of this equality follows directly from property 2 given at the beginning of the section. This formula also enables us to describe the structure of the set $conv(X)$ as follows: $conv(X)$ consists of all those elements $x \in E$ which can be represented in the form

$$x = t_1 x_1 + t_2 x_2 + \ldots + t_k x_k$$

where k is a natural number (depending on x), $x_1, x_2, \ldots, x_k$ are some points from X and $t_1, t_2, \ldots, t_k$ are some real numbers such that

$$0 \leq t_1, \quad 0 \leq t_2, \quad \ldots, \quad 0 \leq t_k,$$

$$t_1 + t_2 + \ \ldots \ + t_k = 1.$$

In other words, the set $conv(X)$ consists of all possible convex combinations of points from X.

If our vector space E is finite-dimensional, then the above-mentioned formula for $conv(X)$ can be essentially simplified. More precisely, the following classical statement due to Carathéodory holds true.

Theorem 10. *Let n be a natural number and let E be a vector space with $dim(E) = n$. If X is an arbitrary subset of E, then*

$$conv(X) = \cup_{j \in J} conv(X_j)$$

where $(X_j)_{j \in J}$ denotes the family of all those subsets of X whose cardinalities are less than or equal to $n + 1$.

The Carathéodory statement can be reformulated in a more geometric form. Namely, we have the next result.

Theorem 11. *Let E be a vector space with $dim(E) = n$ and let P be an n-dimensional convex polyhedron in E. Then there exists a finite family $(P_i)_{i \in I}$ of n-dimensional simplices in E, such that:*
1) $P = \cup_{i \in I} P_i$;
2) the interiors of these simplices are pairwise disjoint;
3) for each $i \in I$, the set of vertices of P_i is contained in the set of vertices of P.

Theorem 11 can easily be proved by induction on n. We suggest the reader to carry out the details of the corresponding argument (note that, from the geometric point of view, the result is almost trivial). Obviously, Theorem 10 is an immediate consequence of Theorem 11.

Let us give one useful application of the Carathéodory theorem.

Theorem 12. *If E is a finite-dimensional topological vector space and X is a compact subset of E, then the set $conv(X)$ is compact in E, too.*

Proof. Evidently, we may suppose without loss of generality that E coincides with the space $\mathbf{R}^n$ where $n \in \mathbf{N}$. Let us denote by T the compact convex subset of $\mathbf{R}^{n+1}$ consisting of all those points

$$(t_1, t_2, ..., t_{n+1}) \in \mathbf{R}^{n+1}$$

for which

$$0 \leq t_1, \ 0 \leq t_2, \ \ldots, \ 0 \leq t_{n+1},$$
$$t_1 + t_2 + ... + t_{n+1} = 1.$$

It can easily be observed that T is an n-dimensional simplex in $\mathbf{R}^{n+1}$ (however, we do not need this fact below). Further, the product $T \times X^{n+1}$ is a compact set in the space

$$\mathbf{R}^{n+1} \times \mathbf{R}^{n(n+1)} = \mathbf{R}^{(n+1)^2}.$$

Let us introduce a mapping

$$\phi \ : \ T \times X^{n+1} \longrightarrow \mathbf{R}^n$$

defined by

$$\phi(t_1, t_2, ..., t_{n+1}, x_1, x_2, ..., x_{n+1}) = t_1 x_1 + t_2 x_2 + ... + t_{n+1} x_{n+1}.$$

Clearly, ϕ is continuous. Besides, according to the Carathéodory theorem formulated above, the range of ϕ coincides with the convex hull of X. Thus, $conv(X)$ is the range of a continuous mapping given on a compact space, so $conv(X)$ is compact as well.

The next statement can be established in a similar way.

Theorem 13. *Let E be a topological vector space and let $(X_i)_{i \in I}$ be a finite family of compact convex subsets of E. Then the convex hull of the set $\cup_{i \in I} X_i$ is compact, too.*

We want to finish this preliminary section with the classical definition of a convex real-valued function and with one simple criterion of continuity of such functions.

Definition 10. Let E be a vector space, let Z be a convex subset of E and let f be a mapping from Z into $\mathbf{R}$. We say that f is lower convex (or, simply, convex) if, for any two points x and y from Z and for each $t \in [0,1]$, we have

$$f(tx + (1-t)y) \leq tf(x) + (1-t)f(y).$$

Furthermore, we say that f is strongly lower convex (or, simply, strongly convex) if the above-mentioned inequality is strong for any two distinct points x and y from Z and for each $t \in]0,1[$.

In a similar manner, the notion of an upper convex (respectively, of strongly upper convex) function can be introduced.

Suppose that f is convex on Z and let k be a nonzero natural number. Then an easy induction on k shows us that, for any points $z_1, z_2, ..., z_k$ from Z and for any k-tuple $(t_1, t_2, ..., t_k)$ of positive real numbers satisfying the relation

$$t_1 + t_2 + ... + t_k = 1,$$

the so-called Jensen inequality

$$f(t_1 x_1 + t_2 x_2 + ... + t_k x_k) \leq t_1 f(x_1) + t_2 f(x_2) + ... + t_k f(x_k)$$

holds true. An analogous strong inequality is valid for strongly convex f.

Let us also observe that, for a convex set $X \subset E$, a function

$$f : X \to \mathbf{R}$$

is convex if and only if the set

$$\{(x,t) \in X \times \mathbf{R} : f(x) \leq t\}$$

is a convex subset of the vector product space $E \times \mathbf{R}$.

Theorem 14. *Let E be a topological vector space, let U be a nonempty open convex subset of E and let*

$$f : U \to \mathbf{R}$$

be a convex function. Then the following two assertions are equivalent:

1) f is continuous on U;

2) there exists a nonempty open subset V of U such that the function $f|V$ (i.e. the restriction of f to V) is bounded from above.

We leave an easy proof of this result to the reader. In particular, if g is a linear functional given on a topological vector space E, then g is continuous on E if and only if there exists a neighbourhood V of 0 such that $g|V$ is bounded.

Note that differential properties of convex functions are more delicate (see, for instance, Exercises 16 and 17 below). However, we do not need these properties in our further considerations.

EXERCISES

1. Let P be a subset of the n-dimensional Euclidean space $\mathbf{R}^n$. Show that the following two assertions are equivalent:

a) P is the convex hull of a finite family of points in $\mathbf{R}^n$ which do not lie in an affine hyperplane of $\mathbf{R}^n$;

b) P is bounded, has nonempty interior and can be represented as the intersection of finitely many closed half-spaces of $\mathbf{R}^n$.

In both cases, P is called an n-dimensional polyhedron in $\mathbf{R}^n$.

2. Let P be a polyhedron in $\mathbf{R}^n$ (not necessarily convex). Demonstrate that there exists a finite family $(T_j)_{j \in J}$ of simplices in $\mathbf{R}^n$, such that

$$P = \cup\{T_j : j \in J\}.$$

Also, formulate the precise definition of a vertex of P.

3. Let E be an arbitrary vector space. By using the existence of an algebraic basis in E, show that the vector structure of E can be enriched up to the structure of a normed vector space, i.e. there exists at least one norm on E. Moreover, demonstrate that such a norm can be chosen as induced by some inner product in E.

On the other hand, give an example of a vector space E whose structure cannot be extended to the structure of a complete normed vector space.

4. Let E be a normed vector space. Prove that the following two assertions are equivalent:

a) E is finite-dimensional;

b) each linear functional defined on E is continuous.

5. Let E be a vector space and let G denote the family of all linear functionals on E. Check that there exists a locally convex topology compatible with the vector structure of E and having the property that all functionals from G are continuous with respect to it. The required topology is, in fact, the weak topology $\sigma(E, G)$ (see, e.g., [29]).

6. Let E be a vector space. Demonstrate that the following two statements are equivalent:

a) E is finite-dimensional;

b) there exists a unique (Hausdorff) topology compatible with the vector structure of E.

7. Let E be an arbitrary vector space. We say that a nonempty subset X of E is a cone in E if

$$(\forall x \in X)(\forall t \geq 0)(tx \in X).$$

In this case, 0 is usually called the vertex of X (note that, according to our definition, $0 \in X$). A cone X is called proper if $X \neq E$. A cone X is called convex if it is a convex set in E. Give a characterization of all maximal (with respect to inclusion) proper convex cones in E (examine the closed half-spaces in E determined by all possible vector hyperplanes).

8. Let X and Y be any two nonempty disjoint closed convex subsets of a finite-dimensional topological vector space E. Show that there exists a (closed) affine hyperplane in E separating X and Y. Moreover, show that, in general, such a hyperplane does not strongly separate these sets (on the other hand, cf. Exercise 11 below).

9. Give an example of an infinite-dimensional Banach space E in which there are two nonempty closed convex sets X and Y such that:

a) $X \cap Y = \emptyset$;

b) there exists no closed affine hyperplane in E separating X and Y.

10. Give an example of an infinite-dimensional Banach space E such that the closed unit ball in E possesses no extreme point.

11. Let E be a locally convex topological vector space, let X be a nonempty closed convex subset of E and let Y be a nonempty compact convex subset of E. Demonstrate that if X and Y are disjoint, then there exists a closed affine hyperplane in E strongly separating X and Y.

12. The standard proof of the Krein-Milman theorem can be found, e.g., in [29] or [179]. Specify this proof and check that the result remains true for compact convex sets in any topological vector space E having the property that E^* separates the points of E.

13. Let E be an arbitrary infinite-dimensional normed vector space. Demonstrate that there exists a partition $\{X, Y\}$ of E such that:

a) each of the sets X and Y is convex in E;

b) both X and Y are everywhere dense in E.

More precisely, by using the method of transfinite recursion, give some construction of the above-mentioned sets X and Y. Note, in connection with this result, that if E is a finite-dimensional normed vector space, then such a partition of E does not exist.

14. Let n be a nonzero natural number and let

$$A = (a_{ij})_{1 \le i \le n,\ 1 \le j \le n}$$

be an $(n \times n)$-matrix whose all terms belong to $\mathbf{R}$. We say that A is doubly stochastic if:

1) $a_{ij} \ge 0$ for all natural indices i and j from $[1, n]$;

2) $\sum_{1 \le j \le n} a_{ij} = 1$ for each natural $i \in [1, n]$;

3) $\sum_{1 \le i \le n} a_{ij} = 1$ for each natural $j \in [1, n]$.

Obviously, any $(n \times n)$-matrix over $\mathbf{R}$ may be regarded as a certain point in the Euclidean space

$$E = \mathbf{R}^n \times \mathbf{R}^n \times \ \dots \ \times \mathbf{R}^n = \mathbf{R}^{n^2}.$$

Check that the set $S(n)$ of all doubly stochastic $(n \times n)$-matrices is convex in E. Check also that the same set is closed and bounded in E. Consequently, $S(n)$ turns out to be a compact convex subset of E.

Further, we say that a doubly stochastic $(n \times n)$-matrix A is a permutation $(n \times n)$-matrix if:

a) for each natural $i \in [1, n]$, all terms a_{ij} $(1 \le j \le n)$ are equal to zero except only one of them (which, evidently, equals 1);

b) for each natural $j \in [1, n]$, all terms a_{ij} $(1 \le i \le n)$ are equal to zero except only one of them (which, evidently, equals 1).

Demonstrate that the set of all extreme points of $S(n)$ coincides with the set of all permutation $(n \times n)$-matrices (this interesting result is due to Birkhoff). In particular, the set $S(n)$ turns out to be a convex polyhedron in E.

Moreover, show (by applying the Carathéodory theorem presented in this section) that any doubly stochastic $(n \times n)$-matrix can be represented as a convex combination of $n^2 - 2n + 2$ permutation $(n \times n)$-matrices.

15. Let P be a convex polyhedron in a finite-dimensional topological vector space E and let X be a subset of P. Show that the following two assertions are equivalent:

a) $X = extr(P)$;

b) X is the smallest (with respect to inclusion) subset of P such that, for any linear functional g on E, we have

$$sup(g(P)) \in g(X), \quad inf(g(P)) \in g(X).$$

16. We recall that, according to the classical theorem of real analysis (see, e.g., [154]), any monotone function acting from **R** into **R** is continuous at all points of **R** except countably many of them. Starting with this theorem, demonstrate that if

$$f : \mathbf{R} \to \mathbf{R}$$

is an arbitrary convex function, then the set of all those points $x \in \mathbf{R}$ for which the derivative $f'(x)$ exists is co-countable (i.e. is the complement of a countable subset of **R**).

Moreover, show that if a convex function

$$g : \mathbf{R} \to \mathbf{R}$$

is such that the derivative $g'(x)$ exists for all points x from **R**, then the function

$$g' : \mathbf{R} \to \mathbf{R}$$

is continuous.

Further, according to another classical result of real analysis, due to Lebesgue, any monotone function h acting from **R** into **R** is differentiable almost everywhere, i.e. the set of all those points x from $\mathbf{R}$, at which the derivative $h'(x)$ exists, is the complement of a Lebesgue measure zero subset of $\mathbf{R}$ (see, e.g., [154] or [177]). Starting with this result, prove that if

$$f : \mathbf{R} \to \mathbf{R}$$

is an arbitrary convex function, then the second derivative $f''(x)$ exists for almost all (in the sense of Lebesgue measure) points $x \in \mathbf{R}$. In other words, f is twice differentiable almost everywhere in $\mathbf{R}$.

It should also be noted that the latter result admits an extension to convex functions defined on the space $\mathbf{R}^n$ (or, more generally, on a convex subset of $\mathbf{R}^n$ with nonempty interior). Namely, it turns out that any such a function possesses the full second differential at almost all points of its domain (in this connection, see e.g. [6]). This result is deep and rather difficult to prove.

17. Let f be a convex function acting from $\mathbf{R}^n$ into $\mathbf{R}$. By applying the method of induction on n and the classical Fubini theorem, give a direct proof of the following result: the full first differential of f exists at almost all (in the Lebesgue sense) points of $\mathbf{R}^n$.

Formulate and prove an analogous result in terms of the Baire category.

2. Brunn-Minkowski inequality

This section is basic for our further considerations and is devoted to those convex sets which lie in finite-dimensional topological vector spaces. As mentioned in the previous section, if E is an arbitrary finite-dimensional (Hausdorff) topological vector space, then E is isomorphic to some Euclidean space $\mathbf{R}^n$. So we can assume, without loss of generality, that all convex sets under consideration below are subsets of $\mathbf{R}^n$. In addition, compact convex sets in $\mathbf{R}^n$ are of primary interest to us. Therefore, it is reasonable to start our discussion with some elementary metrical properties of such sets.

First, let us recall that the compactness of a set $X \subset \mathbf{R}^n$ is equivalent to its closedness and boundedness in $\mathbf{R}^n$. For any two nonempty compact subsets X and Y of $\mathbf{R}^n$, we define

$$d(X, Y) = inf\{r > 0 \ : \ X \subset U_r(Y) \ and \ Y \subset U_r(X)\}$$

where $U_r(X)$ (respectively, $U_r(Y)$) denotes the r-neighbourhood of X (respectively, of Y). The following simple statement is true.

Lemma 1. *The function d is a metric on the family $Comp(\mathbf{R}^n)$ of all nonempty compact subsets of $\mathbf{R}^n$.*

The above-mentioned metric d is usually called the Hausdorff metric on $Comp(\mathbf{R}^n)$. It possesses rather nice properties. In particular, we have

Theorem 1. *The pair $(Comp(\mathbf{R}^n), d)$ is a complete separable metric space.*

Besides, this space turns out to be locally compact. More precisely, we have the following classical result due to Blaschke (see, e.g., [18]).

Theorem 2. *Let $(X_i)_{i \in I}$ be an arbitrary family of nonempty compact subsets of $\mathbf{R}^n$, such that all these sets are contained in some ball of $\mathbf{R}^n$. Then there exists a countable subfamily of $(X_i)_{i \in I}$ converging (in metric d) to some nonempty compact subset of $\mathbf{R}^n$.*

The result just formulated is usually called the Blaschke selection theorem. It follows at once from this result that any closed ball in the space of all nonempty compact subsets of $\mathbf{R}^n$ (equipped with the above-mentioned metric d) is a compact space.

The proofs of Theorems 1 and 2 can be carried out by using the standard diagonalization argument and are not connected with any difficulties (cf. [107], [131]).

Furthermore, it can easily be observed that the limit (in the same metric d) of a sequence of nonempty compact convex sets in $\mathbf{R}^n$ is convex, too. In other words, the class of all nonempty compact convex subsets of $\mathbf{R}^n$ may be regarded as a closed subspace of the class of all nonempty compact subsets of $\mathbf{R}^n$. Therefore, the class of all nonempty compact convex subsets of $\mathbf{R}^n$ turns out to be a locally compact complete separable metric space.

Note now that the natural algebraic operations in $\mathbf{R}^n$ possess canonical extensions to the class of all nonempty compact subsets of $\mathbf{R}^n$. Indeed, let X and Y be any two compact subsets of $\mathbf{R}^n$ and let a and b be any two real numbers. Then the set

$$aX + bY = \{ax + by \ : \ x \in X, \ y \in Y\}$$

is compact, too, because it coincides with the range of the continuous mapping

$$(x, y) \to ax + by \qquad (x \in X, \ y \in Y)$$

defined on the compact product space $X \times Y$. In addition, if the given sets X and Y are convex, then the set $aX + bY$ is convex as well. In particular, the vector sum of an arbitrary finite family of convex subsets of $\mathbf{R}^n$ is a convex subset of $\mathbf{R}^n$. So we can conclude that the space of all nonempty compact convex subsets of $\mathbf{R}^n$ is closed under the natural algebraic operations.

It is not hard to demonstrate, for $X \subset \mathbf{R}^n$, the equivalence of the following two assertions:

1) X is convex in $\mathbf{R}^n$;

2) for any reals $a > 0$ and $b > 0$, the equality

$$(a + b)X = aX + bX$$

is fulfilled.

Obviously, this equivalence remains true for an arbitrary vector space E (over $\mathbf{R}$) and for an arbitrary set X lying in E.

Starting our considerations connected with the Brunn-Minkowski inequality, we wish to point out the following circumstance. Every compact convex body in the space $\mathbf{R}^n$ is measurable in the Jordan sense and, clearly, its n-dimensional Jordan measure (simply, its volume) is strictly positive.

We denote by the symbol v_n the standard function of volume in $\mathbf{R}^n$ which is defined on the class of all Jordan measurable subsets of $\mathbf{R}^n$ and, in particular, on the family of all compact convex bodies in $\mathbf{R}^n$. Evidently, v_n is the restriction of the standard n-dimensional Lebesgue measure λ_n to the class of all Jordan measurable subsets of $\mathbf{R}^n$. For the sake of simplicity, we shall write v, instead of v_n, in all those cases when the dimension n of our Euclidean space is fixed (and no confusion can arise). The function v is finitely additive and invariant under the group of all motions (i.e. isometric transformations) of the Euclidean space.

Now, we are able to formulate the classical Brunn-Minkowski theorem (see [33] and [152]).

Theorem 3. *Let $0 < t < 1$ and let H_0 and H_1 be any two compact convex bodies in the space $\mathbf{R}^n$. Consider the compact convex body*

$$H_t = (1-t)H_0 + tH_1.$$

Then the following inequality holds true:

$$(v(H_t))^{1/n} \geq (1-t)(v(H_0))^{1/n} + t(v(H_1))^{1/n}.$$

In addition, this inequality is reduced to the equality if and only if our bodies H_0 and H_1 are homothetic (the corresponding homothety is assumed to be with a positive coefficient).

First of all, let us point out some immediate consequences of Theorem 3.

1. The real-valued function

$$t \to (v(H_t))^{1/n} \qquad (0 \leq t \leq 1)$$

turns out to be upper convex. It is linear if and only if the original bodies H_0 and H_1 are homothetic.

2. Let X and Y be any two compact convex bodies in $\mathbf{R}^n$ and let a and b be arbitrary strictly positive reals. Then the inequality

$$(v(aX + bY))^{1/n} \geq a(v(X))^{1/n} + b(v(Y))^{1/n}$$

holds true. Furthermore, this inequality is reduced to the equality if and only if X and Y are homothetic.

In order to establish the Brunn-Minkowski theorem (i.e. Theorem 3) it suffices to demonstrate the following auxiliary statement.

Lemma 2. *Let $0 < t < 1$ and let C_0 and C_1 be two compact convex bodies in the space $\mathbf{R}^n$, such that*

$$v(C_0) = v(C_1) = 1.$$

Then, for the compact convex body

$$C_t = (1-t)C_0 + tC_1,$$

the inequality

$$v(C_t) \geq 1$$

holds true. Furthermore, the equality

$$v(C_t) = 1$$

is fulfilled if and only if C_0 is a translate of C_1 (i.e. $C_0 = h(C_1)$ for some translation h of $\mathbf{R}^n$).

Let us show that Lemma 2 immediately implies Theorem 3. Indeed, suppose that the assertion of this lemma is true and take any two compact convex bodies H_0 and H_1 in the space $\mathbf{R}^n$. Then fix any t from the open interval $]0,1[$ and put

$$C_0 = k_0 H_0, \qquad C_1 = k_1 H_1,$$

where

$$k_0 = (1/v(H_0))^{1/n}, \qquad k_1 = (1/v(H_1))^{1/n}.$$

Obviously,

$$v(C_0) = v(C_1) = 1,$$

so we may apply Lemma 2 to the bodies C_0 and C_1. Applying this lemma, we get

$$v(C_\theta) \geq 1$$

where

$$0 < \theta < 1, \qquad C_\theta = (1-\theta)C_0 + \theta C_1.$$

Since θ changes arbitrarily in $]0,1[$, we can put

$$\theta = \frac{t(v(H_1))^{1/n}}{(1-t)(v(H_0))^{1/n} + t(v(H_1))^{1/n}}.$$

Then we obtain

$$C_\theta = \frac{H_t}{(1-t)(v(H_0))^{1/n} + t(v(H_1))^{1/n}},$$

and, consequently,

$$(v(H_t))^{1/n} \geq (1-t)(v(H_0))^{1/n} + t(v(H_1))^{1/n}.$$

The equality appears in the last formula if and only if C_0 is a translate of C_1. But this means that the original bodies H_0 and H_1 are homothetic. We thus conclude that the assertion of Theorem 3 is implied by Lemma 2.

It remains to prove Lemma 2. In order to do it, we use the induction on n. The case when $n \leq 1$ is trivial. Suppose now that $n > 1$ and that the assertion of the lemma has already been established for $n-1$. Take any two compact convex bodies C_0 and C_1 in $\mathbf{R}^n$ with

$$v(C_0) = v(C_1) = 1$$

and fix $t \in]0,1[$. Utilizing a suitable translation of C_0 (if necessary), we may assume without loss of generality that there exists an affine hyperplane L in $\mathbf{R}^n$ support for both these bodies and such that the bodies lie in one half-space determined by L. In all other respects, L can be chosen arbitrarily. In particular, we may arbitrarily choose the direction of a normal to L (this circumstance will be applied below). Further, for each real $\tau \in [0,1]$, there exist unique hyperplanes $L_0(\tau)$ and $L_1(\tau)$ such that:

1) these two hyperplanes are parallel to L;

2) the part of C_0 contained between L and $L_0(\tau)$ has volume τ;

3) the part of C_1 contained between L and $L_1(\tau)$ has volume τ.

Let us put:

$G_0(\tau) = C_0 \cap L_0(\tau)$;

$G_1(\tau) = C_1 \cap L_1(\tau)$;

$a_0(\tau)$ = the distance between L and $L_0(\tau)$;

$a_1(\tau)$ = the distance between L and $L_1(\tau)$;

$s_0(\tau)$ = the $(n-1)$-dimensional volume of $G_0(\tau)$;

$s_1(\tau)$ = the $(n-1)$-dimensional volume of $G_1(\tau)$.

It is clear that

$$d\tau = s_0 da_0, \quad d\tau = s_1 da_1.$$

Consequently, we may write

$$\frac{da_0}{d\tau} = \frac{1}{s_0(\tau)}, \quad \frac{da_1}{d\tau} = \frac{1}{s_1(\tau)}.$$

Now, for any $\tau \in [0,1]$, consider the set

$$G_t(\tau) = (1-t)G_0(\tau) + tG_1(\tau).$$

Evidently, this set is contained in the body C_t and, simultaneously, lies in the hyperplane $L_t(\tau)$ which is parallel to L and whose distance from L is

$$a_t(\tau) = (1-t)a_0(\tau) + ta_1(\tau).$$

Denoting by $s_t(\tau)$ the $(n-1)$-dimensional volume of $G_t(\tau)$ and applying the simplest version of the Fubini theorem, we get

$$v(C_t) \geq \int_0^1 s_t(\tau)da_t(\tau) = \int_0^1 s_t(\tau)a_t'(\tau)d\tau.$$

On the other hand, we have

$$a_t'(\tau) = \frac{1-t}{s_0(\tau)} + \frac{t}{s_1(\tau)}.$$

In view of the inductive assumption, Lemma 2 and Theorem 3 hold true for $n-1$. Hence

$$(s_t(\tau))^{1/(n-1)} \geq (1-t)(s_0(\tau))^{1/(n-1)} + t(s_1(\tau))^{1/(n-1)}.$$

So the preceding relations imply the inequality

$$v(C_t) \geq$$

$$\int_0^1 ((1-t)(s_0(\tau))^{1/(n-1)} + t(s_1(\tau))^{1/(n-1)})^{n-1}((1-t)/s_0(\tau) + t/s_1(\tau))d\tau.$$

Let us demonstrate that the function f under the integral above is always greater than or equal to 1. For this purpose, let us denote

$$\xi = \frac{s_1(\tau)}{s_0(\tau)}.$$

Then an easy calculation shows that f, regarded as a function of ξ, is of the form

$$f(\xi) = (1-t+t\cdot\xi^{1/(n-1)})^{n-1}(1-t+t/\xi).$$

Obviously,

$$lim_{\xi\to0+}f(\xi) = lim_{\xi\to+\infty}f(\xi) = +\infty.$$

Therefore, our f attains its infimum at some point of the interval $]0,+\infty[$. The differentiation of f easily leads to the expression

$$f'(\xi) = \frac{t(1-t)}{\xi^2}(1-t+t\cdot\xi^{1/(n-1)})^{n-2}(\xi^{n/(n-1)}-1).$$

This formula shows that $f'(\xi) = 0$ if and only if $\xi = 1$. In other words, f attains its infimum only at $\xi = 1$. But, for $\xi = 1$, we obviously have $f(\xi) = 1$. Consequently,

$$f(\xi) \geq 1 \qquad (0 < \xi < +\infty)$$

and the sign of equality is only for $\xi = 1$. In this way, we have proved that

$$v(C_t) \geq \int_0^1 1 d\tau = 1.$$

Suppose now that

$$v(C_t) = 1.$$

Then, taking into account the fact that our f is continuous, we infer that

$$s_0(\tau) = s_1(\tau)$$

for all $\tau \in [0, 1]$, which immediately implies

$$a_0(\tau) = a_1(\tau)$$

for all $\tau \in [0, 1]$. Hence we may write

$$\int_0^1 a_0(\tau)d\tau = \int_0^1 a_1(\tau)d\tau.$$

The last formula means that the geometric centre of mass of C_0 and the geometric centre of mass of C_1 are equidistant from the hyperplane L. Evidently, we can find two translations h_0 and h_1 of the space $\mathbf{R}^n$, such that the centers of $h_0(C_0)$ and $h_1(C_1)$ coincide with the origin of $\mathbf{R}^n$. Then the result obtained above shows that, for each unit vector e in the space $\mathbf{R}^n$, the support hyperplanes of $h_0(C_0)$ and $h_1(C_1)$, corresponding to the exterior normal e, coincide. Therefore

$$h_0(C_0) = h_1(C_1),$$

which obviously yields that C_0 is a translate of C_1.

Lemma 2 and Theorem 3 have thus been proved.

It should be noted that the Brunn-Minkowski inequality was generalized for arbitrary Lebesgue measurable subsets of an Euclidean space (see [142]) and was even extended to those sets which are not assumed to be Lebesgue measurable (in this connection, see [91]). However, in this book we do not need such strong versions of Theorem 3.

In our further considerations, we shall meet a number of applications of the Brunn-Minkowski inequality (as mentioned in Preface, applications to probability theory will be of especial interest to us). But first we are going to deduce from Theorem 3 some other useful inequalities.

Lemma 3. *Let X and Y be any two compact convex bodies in the space $\mathbf{R}^n$ and let a and b be strictly positive real numbers. Then the function Φ defined by the formula*

$$\Phi(a, b) = v(aX + bY)$$

turns out to be a homogeneous polynomial of degree n with respect to the variables a and b, i.e. we have

$$v(aX + bY) = \sum_{0 \leq k \leq n} \frac{n!}{(n-k)!k!} a^{n-k} b^k v_{n-k,k}$$

where $v_{n-k,k}$ are some coefficients depending only on X and Y (which are usually called the mixed volumes of X and Y). In addition, we have

$$v_{n,0} = v(X), \qquad v_{0,n} = v(Y).$$

Proof. First, let us consider the case when X and Y are convex polyhedra in $\mathbf{R}^n$. In this case, it is sensible to apply the method of induction on n. For $n \leq 1$, there is nothing to prove. Suppose now that our assertion is valid for $n - 1$ and let us establish it for n. We may assume without loss of generality that the origin of $\mathbf{R}^n$ lies in the interior of the compact convex body

$$Z = aX + bY.$$

Observe that Z is a polyhedron, too. This fact can easily be checked with the aid of the notion of an extreme point. Indeed, each extreme point z of Z admits a representation $z = ax + by$ where x and y are some extreme points of X and Y respectively. The preceding representation immediately implies that the set of all extreme points of Z is finite, i.e. Z is a polyhedron. Now, let us denote by

$$(Z_1, Z_2, ..., Z_r)$$

the injective family of all $(n-1)$-dimensional faces of Z and let, for any natural index $i \in [1, r]$, the symbol h_i denote the distance of 0 from the hyperplane containing Z_i. Then it is clear that

$$v(Z) = (1/n) \sum_{1 \leq i \leq r} h_i s(Z_i)$$

where $s(Z_i)$ is the $(n-1)$-dimensional volume of Z_i. It can easily be verified that all h_i $(1 \leq i \leq r)$ are homogeneous polynomials of degree 1 with respect to a and b. Also, according to our inductive assumption, all $s(Z_i)$ $(1 \leq i \leq r)$ are homogeneous polynomials of degree $n-1$ with respect to a and b. Consequently, $v(Z)$ turns out to be a homogeneous polynomial of degree n with respect to the same variables.

Finally, let us consider the general case when X and Y are arbitrary compact convex bodies in $\mathbf{R}^n$. In this case, there exist two sequences of convex polyhedra

$$\{X_j \ : \ j = 1, 2, ...\}, \qquad \{Y_j \ : \ j = 1, 2, ...\}$$

converging (in the Hausdorff metric d) to the sets X and Y respectively. Then it is obvious that the sequence of convex polyhedra

$$\{aX_j + bY_j \ : \ j = 1, 2, ...\}$$

converges (in the same metric d) to the set $aX + bY$. Also, we have the equalities

$$lim_{j \to \infty} v(aX_j + bY_j) = v(aX + bY),$$

$$lim_{j \to \infty} v(X_j) = v(X),$$

$$lim_{j \to \infty} v(Y_j) = v(Y).$$

But the limit of a sequence of homogeneous polynomials of degree $\leq n$ (with respect to a and b) is a homogeneous polynomial of degree $\leq n$ (with respect to a and b). So we can conclude that the assertion of Lemma 3 is true for any two compact convex bodies X and Y in $\mathbf{R}^n$.

The properties of mixed volumes

$$v_{n-k,k} = v_{n-k,k}(X, Y)$$

play an important role in various questions of geometry, convex analysis, combinatorics and other domains of mathematics (cf. [2], [3], [4], [5], [64], [70], [85]). Sometimes, it is convenient to denote these real-valued characteristics of convex bodies by

$$v(X, X, ..., X, Y, Y, ..., Y) = v_{n-k,k}$$

where $n-k$ copies of X and k copies of Y participate. In particular, it can easily be observed that

$$v_{n-k,k}(X, Y) = v_{k,n-k}(Y, X).$$

In the sequel, only the properties of $v_{1,n-1}$ will be essentially utilized.

Let us formulate and prove the next auxiliary statement which will be useful below.

Lemma 4. *Let M and P be two n-dimensional convex polyhedra in the space $\mathbf{R}^n$, with equal numbers of $(n-1)$-dimensional faces. Denote by*

$$(M_1, M_2, ..., M_r)$$

the injective family of all $(n-1)$-dimensional faces of M and by

$$(P_1, P_2, ..., P_r)$$

the injective family of all $(n-1)$-dimensional faces of P. Suppose that, for each natural index $i \in [1, r]$, the faces M_i and P_i have the same exterior normal unit vector e_i. Suppose also (without loss of generality) that the origin of $\mathbf{R}^n$ lies in the interiors of M and P, and denote by h_i the distance between 0 and the hyperplane containing M_i. Finally, let $s(P_i)$ be, as usual, the $(n-1)$-dimensional volume of P_i. Then the relation

$$v_{1,n-1}(M, P) = v(M, P, P, ..., P) = (1/n) \sum_{1 \leq i \leq r} h_i s(P_i)$$

is satisfied.

Proof. Let $c > 0$. Consider the convex polyhedron

$$T(c) = P + cM.$$

It is easily seen that the sets

$$T_i(c) = P_i + cM_i \qquad (1 \leq i \leq r)$$

are $(n-1)$-dimensional faces of $T(c)$. The distance between 0 and the hyperplane $G_i(c)$ containing the face $T_i(c)$ is

$$a_i = h'_i + ch_i$$

where h'_i denotes the distance between 0 and the hyperplane containing the face P_i. Let us consider the convex polyhedron $Z = Z(c)$ determined by the family of hyperplanes $G_i(c)$ $(1 \leq i \leq r)$. Evidently,

$$T(c) \subset Z(c).$$

The volume of $Z(c)$ can be regarded as a function of variables

$$(a_1, a_2, ..., a_r),$$

because the vectors

$$(e_1, e_2, ..., e_r)$$

are fixed according to the assumption. At the same time, the volume of $Z(c)$ is a function of c. So we may write

$$\frac{dv(Z(c))}{dc} = \sum_{1 \leq i \leq r} \frac{\partial v(Z)}{\partial a_i} \cdot \frac{da_i}{dc}.$$

But it is obvious that

$$\frac{da_i}{dc} = h_i \qquad (1 \leq i \leq r).$$

Also, it can easily be checked that

$$\frac{\partial v(Z)}{\partial a_i} = s(Z_i) \qquad (1 \leq i \leq r),$$

where Z_i denotes the $(n-1)$-dimensional face of $Z = Z(c)$ contained in the hyperplane $G_i(c)$, and $s(Z_i)$ is the $(n-1)$-dimensional volume of Z_i. Now, if c tends to zero, the polyhedron $Z(c)$ tends to the polyhedron P and all the faces Z_i tend to the faces P_i respectively. Consequently, we get

$$\left(\frac{dv(Z(c))}{dc}\right)_{c=0} = \sum_{1 \leq i \leq r} h_i s(P_i).$$

On the other hand, from the purely geometric viewpoint it is clear that

$$v(Z(c)) = v(T(c)) + o(c)$$

and, in virtue of Lemma 3, we have

$$v(T(c)) = v(P) + nc\, v(M, P, P, ..., P) + o(c).$$

This immediately yields

$$\left(\frac{dv(Z(c))}{dc}\right)_{c=0} = lim_{c \to 0+} \frac{v(Z(c)) - v(P)}{c} = nv(M, P, P, ..., P).$$

So we come to the required equality

$$v(M, P, P, ..., P) = (1/n) \sum_{1 \leq i \leq r} h_i s(P_i),$$

and the proof of Lemma 4 has thus been completed.

Actually, the preceding lemma enables us to obtain the precise expression of the mixed volume $v_{1,n-1}$ in that particular case when two convex polyhedra are given in $\mathbf{R}^n$.

Now, our goal is to infer from the Brunn-Minkowski inequality the well-known Minkowski inequality concerning mixed volumes of convex bodies in Euclidean spaces. The Minkowski inequality will be applied later on, in some classical isoperimetric problems (see Section 4).

Theorem 4. *Let H_0 and H_1 be any two compact convex bodies in $\mathbf{R}^n$. Then we have the relation*

$$(v(H_1, H_0, H_0, \ldots H_0))^n \geq v(H_1) \cdot (v(H_0))^{n-1},$$

and the sign of equality participates in this relation if and only if the given bodies are homothetic.

Proof. Let $0 \leq t \leq 1$. Consider the compact convex body

$$H_t = (1-t)H_0 + tH_1.$$

As mentioned earlier, the Brunn-Minkowski inequality implies that the function

$$t \to (v(H_t))^{1/n}$$

is upper convex on the segment $[0,1]$. Consequently, the function f defined on $[0,1]$ by the formula

$$f(t) = (v(H_t))^{1/n} - (1-t)(v(H_0))^{1/n} - t(v(H_1))^{1/n}$$

is upper convex, too (being a sum of three upper convex functions). Furthermore, according to the Brunn-Minkowski inequality, we have

$$f(t) \geq 0 \qquad (0 \leq t \leq 1)$$

and, in addition,

$$f(0) = f(1) = 0.$$

It follows from these facts that $f'(0) \geq 0$ and the equality $f'(0) = 0$ is valid if and only if

$$f(t) = 0 \qquad (0 \leq t \leq 1).$$

Now, let us calculate $f'(0)$ by starting with the formula

$$v(H_t) = \sum_{0 \leq k \leq n} (1-t)^{n-k} t^k \frac{n!}{(n-k)!k!} w_k$$

where, for the sake of simplicity, we put

$$w_k = v_{n-k,k}(H_0, H_1) = v_{k,n-k}(H_1, H_0).$$

After easy calculations (left to the reader) we come to the expression

$$f'(0) = \frac{w_1 - w_0}{w_0^{(n-1)/n}} + w_0^{1/n} - w_n^{1/n}.$$

Combining this expression with $f'(0) \geq 0$, we obtain the desired inequality

$$w_1^n \geq w_n \cdot w_0^{n-1}.$$

Clearly, the sign of equality is possible here if and only if the function f is identically zero. But the latter means that the given bodies H_0 and H_1 are homothetic. This completes the proof of Theorem 4.

As mentioned above, the Minkowski inequality will be helpful in the sequel for establishing some classical isoperimetric inequalities.

Concluding this section, we want to present an integral inequality (due to Anderson [8]) which has numerous applications in probability theory and plays one of the main roles in our further considerations. This inequality will be directly deduced from the Brunn-Minkowski inequality.

Theorem 5. *Let A be a bounded convex body in the space $\mathbf{R}^n$, symmetric with respect to zero, and let*

$$f : \mathbf{R}^n \to \mathbf{R}$$

be a Lebesgue measurable function satisfying the following conditions:

1) for any $x \in \mathbf{R}^n$, we have $f(x) \geq 0$ and $f(-x) = f(x)$;

2) for each strictly positive real θ, the set

$$B_\theta = \{x \in \mathbf{R}^n : f(x) \geq \theta\}$$

is convex in $\mathbf{R}^n$.

Then, for any $t \in [0,1]$ and for any $y \in \mathbf{R}^n$, the relation

$$\int_A f(x + ty)dx \geq \int_A f(x + y)dx$$

holds true.

Proof. In fact, we have to show that

$$\int_{A+ty} f(x)dx \geq \int_{A+y} f(x)dx.$$

Taking arbitrarily θ from $]0, +\infty[$, we put

$$h(\theta) = v((A + ty) \cap B_\theta),$$

$$h^*(\theta) = v((A + y) \cap B_\theta).$$

Let us demonstrate that

$$h(\theta) \geq h^*(\theta) \qquad (0 < \theta < +\infty).$$

For this purpose, define $r = (t + 1)/2$. Then it can easily be checked that $1/2 \leq r \leq 1$ and the inclusion

$$r((A + y) \cap B_\theta) + (1 - r)((A - y) \cap B_\theta) \subset (A + ty) \cap B_\theta$$

is fulfilled. Besides, since the sets A and B_θ are symmetric with respect to zero, we have

$$(A + y) \cap B_\theta = -((A - y) \cap B_\theta).$$

Applying the Brunn-Minkowski inequality to the set

$$C_\theta = r((A + y) \cap B_\theta) + (1 - r)((A - y) \cap B_\theta)$$

and taking into account the fact that

$$v((A + y) \cap B_\theta) = v((A - y) \cap B_\theta),$$

we obtain

$$v(C_\theta) \geq v((A + y) \cap B_\theta).$$

But

$$v((A + ty) \cap B_\theta) \geq v(C_\theta),$$

and we come to the desired relation

$$v((A + ty) \cap B_\theta) \geq v((A + y) \cap B_\theta).$$

Now, we are able to establish the inequality

$$\int_{A+ty} f(x)dx \geq \int_{A+y} f(x)dx.$$

Note first that if we have

$$\int_{A+ty} f(x)dx = +\infty,$$

then there is nothing to prove. So we may suppose that

$$\int_{A+ty} f(x)dx < +\infty.$$

Utilizing the well-known formula from mathematical analysis, we can write

$$\int_{A+ty} f(x)dx = -\int_{0}^{+\infty} \theta dh(\theta),$$

$$\int_{A+y} f(x)dx = -\int_{0}^{+\infty} \theta dh^*(\theta).$$

Integrating by parts and taking into account the relations

$$lim_{\theta \to +\infty} \theta h(\theta) = 0,$$

$$lim_{\theta \to +\infty} \theta h^*(\theta) = 0,$$

we get

$$\int_{A+ty} f(x)dx = \int_{0}^{+\infty} h(\theta)d\theta,$$

$$\int_{A+y} f(x)dx = \int_{0}^{+\infty} h^*(\theta)d\theta.$$

The preceding equalities immediately lead to the required result since, as demonstrated above, we have

$$h(\theta) \geq h^*(\theta) \qquad (0 < \theta < +\infty).$$

This ends the proof of Theorem 5.

EXERCISES

1. Let (E, ρ) be a compact metric space. We denote by $K = K(E)$ the family of all nonempty closed subsets of E and, for any two sets $X \in K$ and $Y \in K$, define

$$d(X,Y) = inf\{r \ : \ r > 0, \ X \subset V_r(Y), \ Y \subset V_r(X)\}$$

where $V_r(X)$ (respectively, $V_r(Y)$) denotes the r-neighbourhood of X (respectively, of Y). Show that the pair (K, d) is a compact metric space.

Formulate and prove the converse assertion.

2. Check that any compact convex body in $\mathbf{R}^n$ is measurable in the Jordan sense (i.e. has Jordan volume). Also, give an example of a compact body in $\mathbf{R}^n$ ($n \geq 2$) which is homeomorphic to the closed unit ball of $\mathbf{R}^n$ but is not measurable in the Jordan sense.

Recall, in this connection, that each compact subset of $\mathbf{R}^n$ is measurable in the Lebesgue sense (and, obviously, has finite measure).

3. Generalize the Brunn-Minkowski inequality to the case of arbitrary compact subsets of $\mathbf{R}^n$. More precisely, demonstrate that if X and Y are any two nonempty compact subsets of $\mathbf{R}^n$ and a and b are any two strictly positive reals, then

$$(\lambda_n(aX + bY))^{1/n} \geq a(\lambda_n(X))^{1/n} + b(\lambda_n(Y))^{1/n}$$

where λ_n denotes, as usual, the standard n-dimensional Lebesgue measure on $\mathbf{R}^n$.

4. Describe all those cases where the inequality of the preceding exercise is reduced to the equality.

5. Let $X_1, X_2, ..., X_m$ be some compact convex bodies in the space $\mathbf{R}^n$ and let $a_1, a_2, ..., a_m$ be some strictly positive reals. Show that the function Φ defined by

$$\Phi(a_1, a_2, ..., a_m) = v(a_1X_1 + a_2X_2 + ... + a_mX_m)$$

is the restriction of a certain homogeneous polynomial of order n. Starting with this result, introduce the mixed volumes for $X_1, X_2, ..., X_m$.

6. Let M and P be two convex polygons on the Euclidean plane $\mathbf{R}^2$. Check directly that the equality

$$v(M, P) = v(P, M)$$

holds true. In addition to this, give a purely geometric interpretation of the above equality.

7. Complete the proof of Lemma 4. Namely, give a detailed explanation of the relation

$$v(Z(c)) = v(T(c)) + o(c)$$

which participates in the proof of Lemma 4.

8. Show that the Minkowski inequality (i.e. Theorem 4 of this section) logically implies the Brunn-Minkowski inequality.

9. Find necessary and sufficient conditions under which the Anderson inequality (see Theorem 5) is reduced to the equality.

10. We recall that a boundary point x of a convex body X in a topological vector space E is a point of strong convexity of X if there exists a hyperplane G in E which is support for X and satisfies the relation $G \cap X = \{x\}$. A convex body in E is called strongly convex if all its boundary points are points of strong convexity.

Let E be a finite-dimensional topological vector space and let X be a compact convex body in E. As usual, we denote by d the Hausdorff metric in E. Show that, for any real number $\varepsilon > 0$, there exists a compact convex body Y in E satisfying the following conditions:

a) $X \subset Y$;

b) $d(X, Y) < \varepsilon$;

c) Y is strongly convex;

d) Y is smooth (i.e. possesses a tangential hyperplane at each its boundary point).

Deduce from this fact that there exists a sequence $(Y_n)_{n \in \mathbf{N}}$ of compact convex bodies in E, for which we have:

e) $X \subset Y_n$ for all natural numbers n;

f) the family $(Y_n)_{n \in \mathbf{N}}$ is decreasing with respect to inclusion;

g) $lim_{n \to \infty} d(X, Y_n) = 0$;

h) all Y_n are smooth and strongly convex.

In particular, we have the equality

$$X = \cap_{n \in \mathbf{N}} Y_n.$$

The result of this exercise will be essentially applied in our further considerations (see, e.g., Section 10).

11. Let f be a function acting from $\mathbf{R}$ into $\mathbf{R}$. We say that f satisfies the weak form of the Jensen inequality (or that f is midpoint convex) if

$$f(\frac{x+y}{2}) \leq \frac{f(x) + f(y)}{2}$$

for all $x \in \mathbf{R}$ and $y \in \mathbf{R}$. Evidently, every function (from $\mathbf{R}$ into $\mathbf{R}$) convex in the usual sense turns out to be midpoint convex. Suppose now that

$$f : \mathbf{R} \to \mathbf{R}$$

is a midpoint convex function for which there exists a nonempty open subinterval of $\mathbf{R}$ such that the restriction of f to this subinterval is bounded from above. Show that, in such a case, f is continuous everywhere and hence is a convex function in the usual sense (i.e. satisfies the Jensen inequality).

Demonstrate also the following more deep result due to Sierpiński: let $g : \mathbf{R} \to \mathbf{R}$ be a midpoint convex function measurable with respect to the standard Lebesgue measure λ_1 on $\mathbf{R}$; then g is convex in the usual sense.

Formulate and prove an analogous result for midpoint convex functions having the Baire property (the definition of the Baire property is given in Appendix 3; for more detailed information about this property, see e.g. [131] and [159]).

By starting with the general theorem from the theory of vector spaces which states that any vector space (over an arbitrary field) possesses a basis, show that there exists a basis of $\mathbf{R}$ regarded as a vector space over the field $\mathbf{Q}$ of all rational numbers (the so-called Hamel basis). Infer from this fact that there exists a bijective function $h : \mathbf{R} \to \mathbf{R}$ satisfying the relation

$$h(x + y) = h(x) + h(y),$$

for all $x \in \mathbf{R}$ and $y \in \mathbf{R}$, and having the property that its graph is an everywhere dense subset of the plane $\mathbf{R}^2$. In other words, h may be regarded as an automorphism of the additive group $\mathbf{R}$, whose graph is dense in $\mathbf{R}^2$. Consequently, h is not measurable in the Lebesgue sense and does not possess the Baire property. Conclude from this result that h is midpoint convex but is not convex.

3. Convex polyhedra

As known, Minkowski was one of the founders of modern theory of convex sets. He established many deep and important facts concerning such sets in a finite-dimensional Euclidean space. In particular. several beautiful results are due to him in connection with properties of convex polyhedra.

Here we wish to discuss some properties of convex polyhedra lying in the space $\mathbf{R}^n$. Especially, we are interested in the Minkowski classical theorem on the existence and uniqueness of a convex polyhedron with given directions of its exterior normals and given $(n-1)$-dimensional volumes of its $(n-1)$-dimensional faces (briefly, $(n-1)$-faces).

We shall present below the precise formulation of the above-mentioned Minkowski theorem. For this purpose, we begin with some simple statements about exterior normals of a given convex polyhedron.

Let M be a convex polyhedron in the space $\mathbf{R}^n$ where $n \geq 2$. Let us denote by

$$(M_1,\ M_2,\ ...,\ M_k)$$

the injective family of all $(n-1)$-faces of M. We equip each M_i, where $i = 1, 2, ..., k$, with its exterior normal unit vector e_i. This means that:

a) the length of e_i is equal to 1;

b) e_i is orthogonal to the hyperplane G_i containing the face M_i;

c) e_i is directed into that half-space determined by G_i which does not intersect the interior of M.

In other words, e_i is an exterior normal to the polyhedron M at its face M_i $(i = 1, 2, ..., k)$.

Let us denote by s the function of the standard $(n-1)$-dimensional volume in $\mathbf{R}^n$ (actually, s plays the role of the function v_{n-1}). Then it is not hard to verify that

$$s(M_1)e_1 + s(M_2)e_2 + ... + s(M_k)e_k = 0,$$

i.e. the sum of vectors on the left-hand side of this equality yields zero. The proof of this fact can be done, for example, in the following manner. Take an arbitrary hyperplane G in $\mathbf{R}^n$ and denote by $e = e_G$ the normal unit vector to G. For each natural index $i \in [1, k]$, the inner product $< e_i, e >$ is equal to the cosine of the angle between e_i and e. Consequently, the number

$$s(M_i) < e_i, e >$$

coincides with the oriented $(n-1)$-dimensional volume of the projection of M_i on G. Actually, this volume is positive if the angle between e_i and e is less than or equal to $\pi/2$, and is negative if the above-mentioned angle is greater than or equal to $\pi/2$. Note now that the usual $(n-1)$-dimensional volume of the projection of M on G coincides with the sum

$$\sum_{i \in I_1} s(M_i) < e_i, e >$$

where I_1 denotes the set of all those natural indices from $[1, k]$ for which we have $< e_i, e > \geq 0$. In a similar way, we see that the same $(n-1)$-dimensional volume coincides with the sum

$$\sum_{i \in I_2} -s(M_i) < e_i, e >$$

where I_2 denotes the set of all those natural indices from $[1, k]$ for which $< e_i, e > < 0$. Since $\{I_1, I_2\}$ forms a partition of $[1, k]$, we may write

$$\sum_{1 \leq i \leq k} s(M_i) < e_i, e > = 0$$

or, equivalently,

$$< \sum_{1 \leq i \leq k} s(M_i) e_i, e > = 0.$$

Keeping in mind the fact that G (and hence e) was taken arbitrarily, we conclude that the vector

$$\sum_{1 \leq i \leq k} s(M_i) e_i$$

must be zero, which yields the required result.

Note that our assumption that the polyhedron M is convex is not essential here: the result remains true for polyhedra homeomorphic to a closed ball in $\mathbf{R}^n$. The reader can check it without any difficulty.

Let us put

$$s_i = s(M_i) \qquad (1 \leq i \leq k).$$

Then, in virtue of the said above, the following relations are satisfied:

1) the vectors e_i $(i = 1, 2, ..., k)$ are pairwise distinct, do not lie in one hyperplane and

$$||e_1|| = ||e_2|| = ... = ||e_k|| = 1;$$

2) all numbers s_i $(i = 1, 2, ..., k)$ are strictly positive;

3) $\sum_{1 \leq i \leq k} s_i e_i = 0$.

Conversely, Minkowski demonstrated that if two finite families

$$(e_i)_{1 \leq i \leq k} \qquad (s_i)_{1 \leq i \leq k}$$

of vectors in $\mathbf{R}^n$ and of reals are given satisfying relations 1) - 3), then there exists a convex polyhedron in $\mathbf{R}^n$ for which all e_i $(i = 1, 2, ..., k)$ are its exterior normal unit vectors and all s_i $(i = 1, 2, ..., k)$ are volumes of its corresponding $(n-1)$-faces.

In other words, conditions 1) - 3) turn out to be necessary and sufficient for the existence of a convex polyhedron in $\mathbf{R}^n$ having the above-mentioned properties in terms of its exterior normals and volumes of $(n-1)$-faces.

We are going to prove this beautiful result here. So, let us assume that 1) - 3) are fulfilled. Pass through the origin of $\mathbf{R}^n$ the half-lines

$$l_1, \ l_2, ..., \ l_k$$

in such a way that every l_i $(i = 1, 2, ..., k)$ has direction e_i and its end-point coincides with 0. Then fix a family

$$(h_1, \ h_2, \ ..., \ h_k)$$

of strictly positive reals and, for each natural $i \in [1, k]$, take the affine hyperplane $G_i = G_i(h_i)$ in $\mathbf{R}^n$ orthogonal to e_i and such that its distance from the origin is equal to h_i (hence the linear segment on l_i between 0 and $G_i(h_i)$ has length h_i). This family of hyperplanes generates some convex polyhedron in $\mathbf{R}^n$. The obtained polyhedron will be denoted by the symbol

$$M = M(h_1, h_2, ..., h_k)$$

since it is completely determined by $(h_1, h_2, ..., h_k)$. The n-dimensional volume of this polyhedron will be denoted by

$$v(M(h_1, h_2, ..., h_k)) = v(h_1, h_2, ..., h_k).$$

In addition, for any natural $i \in [1, k]$, we denote by $M_i = M_i(h_1, h_2, ..., h_k)$ the $(n-1)$-face of M which is contained in the hyperplane G_i. As usual, the symbol $s(M_i)$ denotes the $(n-1)$-dimensional volume of M_i.

The following auxiliary statement is almost trivial from the geometric viewpoint, so we leave its proof to the reader.

Lemma 1. *For each natural number $i \in [1, k]$, the relation*

$$\frac{\partial v(h_1, h_2, ..., h_k)}{\partial h_i} = s(M_i)$$

holds true.

In connection with Lemma 1, let us point out that if there is no $(n-1)$-face of our polyhedron M, contained in the hyperplane G_i, then the right-hand side of the above-mentioned equality is assumed to be equal to zero.

Consider now the subfamily of the family

$$\{M(h_1, h_2, ..., h_k) \ : \ h_1 > 0, \ h_2 > 0, \ ..., \ h_k > 0\},$$

consisting of all those polyhedra $M(h_1, h_2, ..., h_k)$ for which the following relation is fulfilled:

$$\sum_{1 \leq i \leq k} s_i h_i = 1.$$

First of all, we observe that this subfamily is not empty. Indeed, for sufficiently small strictly positive h_i $(i = 1, 2, ..., k)$, we obviously have

$$\sum_{1 \leq i \leq k} s_i h_i < 1.$$

On the other hand, if h_i $(i = 1, 2, ..., k)$ are sufficiently large, then

$$\sum_{1 \leq i \leq k} s_i h_i > 1.$$

So, in virtue of the continuity, there are some strictly positive reals

$$h'_1, \ h'_2, \ ..., \ h'_k$$

for which the equality

$$\sum_{1 \leq i \leq k} s_i h'_i = 1$$

holds true. This shows that the polyhedron $M(h'_1, h'_2, ..., h'_k)$ belongs to our subfamily.

Further, it is clear that all polyhedra from the above-mentioned subfamily are uniformly bounded (i.e. are contained in some fixed bounded subset of $\mathbf{R}^n$). Indeed, this circumstance follows directly from the relations

$$h_1 < 1/s_1, \ h_2 < 1/s_2, \ \ldots, \ h_k < 1/s_k.$$

Therefore, the volumes of these polyhedra are bounded from above. Let us denote by $v^{(0)}$ the supremum of all these volumes. Taking into account the fact that the function

$$(h_1, h_2, ..., h_k) \to v(h_1, h_2, ..., h_k)$$

is continuous, we infer that, for some polyhedron

$$M^{(0)} = M(h_1^{(0)}, h_2^{(0)}, ..., h_k^{(0)})$$

belonging to the closure of our subfamily (with respect to the Hausdorff metric d), the equality

$$v^{(0)} = v(M^{(0)})$$

must be satisfied. We are going to demonstrate that $M^{(0)}$ belongs to our subfamily, as well. For this purpose, observe that the relation

$$\sum_{1 \le i \le k} s_i h_i = 1$$

is preserved in view of the continuity. Hence, for $M^{(0)}$, we have

$$\sum_{1 \le i \le k} s_i h_i^{(0)} = 1.$$

It remains to verify that

$$h_1^{(0)} > 0, \quad h_2^{(0)} > 0, \quad ..., \quad h_k^{(0)} > 0.$$

From the geometric point of view, the last relations mean that the origin of $\mathbf{R}^n$ lies in the interior of $M^{(0)}$. Suppose to the contrary that the origin is not in $int(M^{(0)})$. Then we can choose a vector $e \in \mathbf{R}^n$ having the property that

$$0 \in int(M^{(1)}),$$

where

$$M^{(1)} = M(h_1^{(1)}, h_2^{(1)}, ..., h_k^{(1)}) = e + M^{(0)}.$$

Evidently, we get

$$h_1^{(1)} > 0, \quad h_2^{(1)} > 0, \quad ..., \quad h_k^{(1)} > 0,$$

$$h_i^{(1)} = h_i^{(0)} + < e_i, e > \qquad (1 \le i \le k),$$

$$v(M^{(1)}) = v(M^{(0)}) = v^{(0)},$$

$$\sum_{1\leq i\leq k} s_i h_i^{(1)} = \sum_{1\leq i\leq k} s_i h_i^{(0)} + < \sum_{1\leq i\leq k} s_i e_i, e > = 1 + 0 = 1.$$

So, the polyhedron $M^{(1)}$ can play the role of $M^{(0)}$.

Thus, we may suppose in our further considerations that $M^{(0)}$ belongs to the above-mentioned subfamily of [illegible]ra. In particular, we observe that $M^{(0)}$ yields a solution of a cer[illegible] p[illegible] [illegible]f a relative extremum. More precisely, the function $v = v(h_1, h_2, ..., h_k)$ attains its maximum at the point $(h_1^{(0)}, h_2^{(0)}, ..., h_k^{(0)})$ under the condition that

$$\sum_{1\leq i\leq k} s_i h_i = 1.$$

Then, according to the widely known Lagrange criterion, we must have

$$\frac{\partial}{\partial h_i}(v(h_1, h_2, ..., h_k) - \alpha \sum_{1\leq i\leq k} s_i h_i) = 0 \qquad (i = 1, 2, ..., k)$$

for some nonzero real coefficient α. Calculating partial derivatives and taking into account Lemma 1, we easily come to the equalities

$$s(M_i^{(0)}) = \alpha s_i \qquad (i = 1, 2, ..., k),$$

where, for each natural $i \in [1, k]$, the symbol $M_i^{(0)}$ denotes the $(n-1)$-face of $M^{(0)}$ corresponding to the vector e_i. Finally, we put

$$M = \beta M^{(0)}$$

where the coefficient β is defined by

$$\beta = (1/\alpha)^{1/(n-1)}.$$

It is clear now that, for each natural index $i \in [1, k]$, the convex polyhedron M has an $(n-1)$-dimensional face M_i which is orthogonal to e_i and

$$s(M_i) = s_i.$$

Hence, the family $(e_i)_{1\leq i\leq k}$ consists of exterior normal unit vectors for M and the family $(s_i)_{1\leq i\leq k}$ consists of $(n-1)$-dimensional volumes of $(n-1)$-faces of M. In other words, the polyhedron M is the required one. Thus, we have proved the existence of a convex polyhedron with suitable properties.

Now, let us demonstrate that such a polyhedron M must be unique. We shall show that the uniqueness of M follows at once from the Brunn-Minkowski inequality (namely, from its degenerate case when it is reduced to the equality).

Let any two convex polyhedra M and P be given in the space $\mathbf{R}^n$, where $n \geq 2$, and suppose that they have the same exterior normal unit vectors $(e_i)_{1 \leq i \leq k}$ and the same $(n-1)$-dimensional volumes of $(n-1)$-faces respectively. More precisely, the injective family

$$(M_1, M_2, ..., M_k)$$

of all $(n-1)$-dimensional faces of M and the injective family

$$(P_1, P_2, ..., P_k)$$

of all $(n-1)$-dimensional faces of P are such that, for each natural number $i \in [1, k]$, the vector e_i is an exterior normal to M_i and P_i simultaneously, and

$$s(M_i) = s(P_i).$$

Let us verify that, in this case, M is a translate of P. We may assume, without loss of generality, that the origin of $\mathbf{R}^n$ lies in the interiors of M and P. For any natural $i \in [1, k]$, let us denote by a_i the distance between 0 and the hyperplane containing M_i. Denote also by b_i the distance between 0 and the hyperplane containing P_i. Further, let us recall the formula for the mixed volume, which was established in the previous section (see Lemma 4 therein). In virtue of that formula, we have

$$v(M, P, P, ..., P) = (1/n) \sum_{1 \leq i \leq k} a_i s(P_i).$$

In view of our assumptions,

$$(1/n) \sum_{1 \leq i \leq k} a_i s(P_i) = (1/n) \sum_{1 \leq i \leq k} a_i s(M_i) = v(M).$$

According to the Minkowski inequality,

$$(v(M, P, P, ..., P))^n \geq v(M)(v(P))^{n-1}.$$

Hence we get

$$(v(M))^n \geq v(M)(v(P))^{n-1}$$

and, consequently,

$$v(M) \geq v(P).$$

Interchanging the roles of M and P and applying the same argument, we obtain

$$v(P, M, M, ..., M) = (1/n) \sum_{1 \leq i \leq k} b_i s(M_i) = v(P),$$

$$(v(P))^n \geq v(P)(v(M))^{n-1},$$

$$v(P) \geq v(M).$$

Thus, we can conclude that

$$v(M) = v(P).$$

But then the equality

$$(v(M, P, P, ..., P))^n = v(M)(v(P))^{n-1}$$

must be true. On the other hand, as demonstrated in Section 2, the latter equality is satisfied if and only if the given polyhedra M and P are homothetic. Since these polyhedra have the same volume, we may assert that the homothety coefficient is equal to 1 and therefore M turns out to be a translate of P. This establishes the uniqueness of a polyhedron possessing the properties described.

In this way, we have proved the following classical theorem due to Minkowski (see [150]).

Theorem 1. *Let $(s_i)_{1 \leq i \leq k}$ be a finite family of strictly positive reals and let $(e_i)_{1 \leq i \leq k}$ be an injective family of unit vectors in the space $\mathbf{R}^n$ which do not lie in one hyperplane and satisfy the relation*

$$\sum_{1 \leq i \leq k} s_i e_i = 0.$$

Then there exists a unique (up to translations) convex polyhedron M in $\mathbf{R}^n$ for which the reals s_i coincide with $(n-1)$-dimensional volumes of its $(n-1)$-dimensional faces and the vectors e_i are exterior normals of these faces.

Some consequences and applications of Theorem 1 are presented below in exercises to this section. The reader who is especially interested in the general theory of convex polyhedra can find an extensive interesting material in the classical monograph by Alexandrov [7] where numerous beautiful results from this theory are considered and the corresponding deep techniques is developed. It is reasonable to point out here that essentially another proof of Theorem 1 is given in that monograph, based on some ideas of combinatorial (algebraic) topology.

For many other important facts from the theory of convex polyhedra, we refer the reader to [72], [84], [85], [88], [120] and [143] (see also exercises below).

EXERCISES

1. Give a detailed proof of Lemma 1 from this section.

2. Let E be a topological vector space, let X be a subset of E and let $x \in X$. We say that the set X is locally conic at the point x if there exists a neighbourhood $U(x)$ of this point, such that

$$U(x) \cap X = U(x) \cap K(x),$$

where $K(x)$ is some cone in E with vertex x (i.e. $K(x)$ is a translate of some cone with vertex 0 and x is the image of 0 under that translation).

Now, let E be a locally convex topological vector space and let M be a compact convex subset of E locally conic at each its point belonging to the closure of $extr(M)$. Show, by applying the Krein-Milman theorem, that M is a convex polyhedron in E.

This yields a characterization of convex polyhedra in terms of those points at which a given compact convex set is locally conic.

3. Show that, for a natural number $n \geq 3$, there exists a compact convex body T in the space $\mathbf{R}^n$, satisfying the following conditions:

a) T is locally conic at each its extreme point;

b) T is not a polyhedron in $\mathbf{R}^n$.

On the other hand, verify that if T is a two-dimensional compact convex subset of $\mathbf{R}^2$ locally conic at all its extreme points, then T is a convex polygon in $\mathbf{R}^2$.

4. Let T be a compact convex body in the space $\mathbf{R}^n$ and let $x \in T$. We say that T is locally polyhedral at x if there exists a neighbourhood $U(x)$ of x satisfying the relation

$$U(x) \cap T = U(x) \cap M(x),$$

where $M(x)$ is some convex polyhedron in $\mathbf{R}^n$ containing x.

Demonstrate (by using the induction on n) that the following two assertions are equivalent:

a) T is locally polyhedral at each its extreme point;
b) T is a convex polyhedron in $\mathbf{R}^n$.

5. Let T be a compact convex set in the space $\mathbf{R}^n$ having the property that, for each $x \in \mathbf{R}^n$, the set $(T+x) \cap T$ is either empty, or one-element, or is a homothetic image of T (where the coefficient of homothety is assumed to be positive). Show that, in this case, the body T is a simplex in $\mathbf{R}^n$ whose dimension is less than or equal to n.

The result of Exercise 5 is due to Rogers and Shephard.

6. Let $(T_k)_{k \in \mathbf{N}}$ be a decreasing (with respect to inclusion) sequence of simplices in the space $\mathbf{R}^n$. Show (by applying the result of Exercise 5) that the set

$$T = \cap \{T_k \ : \ k \in \mathbf{N}\}$$

is a simplex in $\mathbf{R}^n$. This result is due to Kolmogorov and Borovikov.

Let $(S_k)_{k \in \mathbf{N}}$ be an increasing (with respect to inclusion) sequence of simplices in $\mathbf{R}^n$, such that the union of all these simplices is a bounded subset of $\mathbf{R}^n$. Show that the set

$$S = cl(\cup \{S_k \ : \ k \in \mathbf{N}\})$$

is a simplex in $\mathbf{R}^n$.

7. We recall that a parallelepiped in $\mathbf{R}^n$ is any compact convex polyhedron P which can be represented in the form

$$P = \{e + t_1 e_1 + t_2 e_2 + \ldots + t_k e_k \ : \ (t_1, t_2, \ldots, t_k) \in [0,1]^k\},$$

where k is a natural number, e, e_1, e_2, ..., e_k are some fixed vectors in $\mathbf{R}^n$ and, in addition, e_1, e_2, ..., e_k are linearly independent. More precisely, P is called a k-dimensional parallelepiped in $\mathbf{R}^n$.

Now, let q be a fixed number from the open interval $]0,1[$ and let T be a nonempty compact convex set in $\mathbf{R}^n$ having the property that, for each $x \in \mathbf{R}^n$, the intersection $(qT + x) \cap T$ is centrally symmetric. Prove that, in such a case, T is a parallelepiped in $\mathbf{R}^n$ (whose dimension is less than or equal to n).

In connection with Exercise 7, see also [109].

8. Let $(P_k)_{k \in \mathbf{N}}$ be a decreasing (with respect to inclusion) sequence of parallelepipeds in $\mathbf{R}^n$. Show (by applying the result of Exercise 7) that the set

$$P = \cap \{P_k \ : \ k \in \mathbf{N}\}$$

is a parallelepiped in $\mathbf{R}^n$. This statement is due to Gruber.

Let $(L_k)_{k \in \mathbf{N}}$ be an increasing (with respect to inclusion) sequence of parallelepipeds in $\mathbf{R}^n$ such that the union of all these parallelepipeds is a bounded subset of $\mathbf{R}^n$. Show that the set

$$L = cl(\cup\{L_k \ : \ k \in \mathbf{N}\})$$

is a parallelepiped in $\mathbf{R}^n$.

9. Let M be a convex polyhedron in the space $\mathbf{R}^n$ which admits a decomposition into finitely many centrally symmetric convex polyhedra. Prove (by utilizing the Minkowski theorem presented in this section) that M is centrally symmetric. Show also that the assumption of convexity of M is essential here.

10. Let T be a compact convex body in $\mathbf{R}^n$ having the property that the orthogonal projection of T on an arbitrary affine hyperplane of $\mathbf{R}^n$ is centrally symmetric. Prove that T is centrally symmetric.

11. Let $n \geq 3$ and let M be a convex n-dimensional polyhedron in $\mathbf{R}^n$ such that all $(n-1)$-faces of M are centrally symmetric. Demonstrate that M is centrally symmetric, too (deduce this fact from the results of Exercises 9 and 10).

12. Let M be a convex n-dimensional polyhedron in $\mathbf{R}^n$. Show that the following two assertions are equivalent:

a) all faces of M are centrally symmetric;

b) M admits a decomposition into finitely many n-dimensional parallelepipeds.

Note that, in assertion a), the faces are assumed to be of all possible dimensions k (where $0 \leq k \leq n$).

13. Let $n \geq 3$ and let T be a compact convex body in $\mathbf{R}^n$ having the property that the orthogonal projection of T on an arbitrary affine hyperplane of $\mathbf{R}^n$ is a convex polyhedron in that hyperplane. Prove that T is a polyhedron in $\mathbf{R}^n$. This result is due to Klee.

14. Generalize the Minkowski theorem presented in this section to arbitrary compact convex bodies in $\mathbf{R}^n$. In order to obtain such a generalization, apply an approximation of any compact convex body in $\mathbf{R}^n$ by convex polyhedra in the same space.

In connection with Exercise 14, see also [7] and [48].

15. Let M be a convex polyhedron in the space $\mathbf{R}^n$ with $dim(M) = n$. Denote by $(M_1, M_2, ..., M_k)$ the injective family of all $(n-1)$-faces of M. Let

M' be another n-dimensional convex polyhedron in $\mathbf{R}^n$ whose $(n-1)$-faces are M'_1, M'_2, ..., M'_k. Suppose also that:

a) M is contained in M';

b) for each natural $i \in [1, k]$, the hyperplane G_i containing M_i is parallel to the hyperplane G'_i containing M'_i;

c) for each natural $i \in [1, k]$, the distance between the hyperplanes G_i and G'_i is equal to 1.

Show that the following two assertions are equivalent:

1) there exists a sphere in the space $\mathbf{R}^n$, contained in M and tangential to all hyperplanes G_i $(i = 1, 2, ..., k)$;

2) the polyhedra M and M' are similar, i.e. there exists a transformation f of the space $\mathbf{R}^n$, representable as a composition of some homothety and motion of $\mathbf{R}^n$, such that $M' = f(M)$.

16. Let S and S' be two n-dimensional simplices in $\mathbf{R}^n$ for which there is a one-to-one correspondence between the families of their $(n-1)$-faces, such that the corresponding $(n-1)$-faces are congruent. Demonstrate that:

a) S and S' have the same radius of circumscribed spheres;

b) the volumes of S and S' are equal;

c) S and S' have the same radius of inscribed spheres.

Find some other corresponding real-valued geometric characteristics of these simplices which also turn out to be equal.

In connection with Exercise 16, let us remark that it is unknown at the present time whether the simplices S and S' satisfying the assumption of this exercise are congruent. Actually, we deal here with a geometric version of the famous Ulam problem from graph theory (see [195]).

4. Two classical isoperimetric problems

We have already mentioned in the Preface that we are not going to discuss in detail various aspects of the theory of isoperimetric problems. This topic is very wide and is under extensive investigations. The reader who is interested in these questions, can be referred to many books and monographs devoted to the subject, e.g., [18], [47], [72], [85], [134] and [165]. However, it is not reasonable to miss some well-known statements from this topic, which are basic for further results and facts of the theory. We shall consider here two classical geometric theorems of isoperimetric type. The choice of these theorems is motivated by the circumstance that both of them are closely connected with the Brunn-Minkowski inequality.

Let us suppose that an injective finite family

$$(e_1, e_2, ..., e_k)$$

of unit vectors in the space $\mathbf{R}^n$, where $n \geq 2$, is given such that there exists no closed half-space containing all of them. Then it is not hard to observe that there are various convex polyhedra in $\mathbf{R}^n$ for which this family turns out to be the family of all exterior normals. Indeed, our vectors belong to the unit sphere of $\mathbf{R}^n$. So we may consider, for each natural index $i \in [1, k]$, the hyperplane G_i tangential to the sphere at the point e_i. In this way, we obtain the family $(G_i)_{1 \leq i \leq k}$ of hyperplanes which determine a certain convex polyhedral set containing the sphere. According to our assumption, this set is bounded in $\mathbf{R}^n$ and hence is a polyhedron. Also, it immediately follows from the construction presented that the family of exterior normals of this polyhedron coincides with $(e_i)_{1 \leq i \leq k}$. It will be convenient to denote the above-mentioned polyhedron by

$$M^{(0)} = M^{(0)}(e_1, e_2, ..., e_k).$$

Our goal is to establish one remarkable property of $M^{(0)}$ discovered by Lindelöf (see [138]).

Let s denote, as earlier, the standard function of $(n-1)$-dimensional volume in $\mathbf{R}^n$ and let

$$(M_1^{(0)}, M_2^{(0)}, ..., M_k^{(0)})$$

be the family of all $(n-1)$-dimensional faces of $M^{(0)}$. The enumeration of faces is chosen so that, for each natural $i \in [1,k]$, the face $M_i^{(0)}$ has exterior normal e_i. Then it is clear that

$$v(M^{(0)}) = (1/n) \sum_{1 \leq i \leq k} s(M_i^{(0)}),$$

because all distances of the origin from faces $M_i^{(0)}$ $(1 \leq i \leq k)$ are equal to 1.

Further, consider in the space $\mathbf{R}^n$ any convex n-dimensional polyhedron M such that the family of all exterior normals of M is the same family of vectors

$$(e_1, e_2, ..., e_k),$$

and denote by

$$(M_1, M_2, ..., M_k)$$

the family of all $(n-1)$-dimensional faces of M (we suppose again that, for each natural $i \in [1,k]$, the face M_i corresponds to the vector e_i, i.e. this vector is an exterior normal of M_i). Applying the formula of mixed volume (see Lemma 4 from Section 2) to $M^{(0)}$ and M, we easily infer that

$$v(M^{(0)}, M, M, ..., M) = (1/n) \sum_{1 \leq i \leq k} s(M_i).$$

Suppose now that our polyhedra $M^{(0)}$ and M have the same surface area, i.e.

$$\sum_{1 \leq i \leq k} s(M_i^{(0)}) = \sum_{1 \leq i \leq k} s(M_i) = a.$$

Then, in virtue of the Minkowski inequality (see Theorem 4 from Section 2), the following relations hold:

$$(a/n)^n = (v(M^{(0)}, M, M, ..., M))^n \geq v(M^{(0)})(v(M))^{n-1},$$

$$(a/n)^n \geq (a/n)(v(M))^{n-1},$$

$$a/n \geq v(M),$$

$$v(M^{(0)}) \geq v(M).$$

In the last relation, the equality is fulfilled if and only if the polyhedra $M^{(0)}$ and M are homothetic. But, since these polyhedra have the same surface area, the homothety coefficient must be equal to 1. Hence, in the case of the equality, $M^{(0)}$ is a translate of M.

In this way, we have proved the following classical statement due to Lindelöf [138].

Theorem 1. *Among all convex polyhedra in* $\mathbf{R}^n$ $(n \geq 2)$ *with given exterior normals and a fixed surface area, a polyhedron circumscribed around a sphere (and only it) has a greatest volume.*

Note that Theorem 1 possesses many other proofs based on essentially different ideas. Moreover, a quite elementary proof of this theorem is possible for $n = 2$ (see Exercise 10 to this section).

Now, let us turn our attention to another classical isoperimetric problem and establish (with the aid of the Minkowski inequality) that, among all compact convex bodies in $\mathbf{R}^n$ having fixed surface area, a ball (and only it) has the greatest volume. This isoperimetric property of a ball has very long history and during centuries stimulated the development of the general theory of extremal problems. In this connection, some interesting historical comments can be found e.g. in [18].

We first need to specify the notion of a surface area for compact convex bodies in the space $\mathbf{R}^n$.

Let A and B be any two compact convex bodies in $\mathbf{R}^n$. For each strictly positive real number r, consider the expression

$$\frac{v(A + rB) - v(A)}{r}$$

where, as usual, v denotes the standard volume function in $\mathbf{R}^n$. Suppose that

$$lim_{r \to 0+} \frac{v(A + rB) - v(A)}{r}$$

does exist for given A and B. Then the value of this limit is called the surface area of A with respect to B (in the sense of Minkowski).

The following auxiliary statement is valid.

Lemma 1. *If B coincides with the closed unit ball in the space* $\mathbf{R}^n$, *then the above-mentioned limit does exist and is equal to the usual surface area of A.*

The proof of Lemma 1 is not difficult and we leave it to the reader as a useful exercise concerning the basic concepts of volume and area. In order to avoid misunderstanding, we only wish to recall here that the surface area of a compact convex body A is usually defined as the supremum of all surface areas of those convex polyhedra which are contained in A.

Let us return to the case of an arbitrary compact convex body B and suppose that, for a compact convex body A, there exists a surface area of A with respect to B. Then, according to the Brunn-Minkowski inequality, we may write

$$v(A + rB) \geq ((v(A))^{1/n} + r(v(B))^{1/n})^n$$

whenever r is a strictly positive real number. Expanding the right-hand side of this inequality, we get

$$v(A + rB) \geq v(A) + nr(v(A))^{(n-1)/n}(v(B))^{1/n} + o(r).$$

Consequently,

$$\frac{v(A + rB) - v(A)}{r} \geq n(v(A))^{(n-1)/n}(v(B))^{1/n} + o(r)/r.$$

Tending r to zero, we obviously obtain

$$s_B(A) \geq n(v(A))^{(n-1)/n}(v(B))^{1/n}$$

where $s_B(A)$ denotes the surface area of A with respect to B.

In particular, if B coincides with the closed unit ball in $\mathbf{R}^n$, then

$$s(A) \geq n(v(A))^{(n-1)/n}(v(B))^{1/n}$$

where $s(A)$ is the usual surface area of A. Assuming again that B is the closed unit ball in $\mathbf{R}^n$, let us denote

$$\beta = n(v(B))^{1/n} = const.$$

Then the preceding inequality can be rewritten in the form

$$s(A) \geq \beta(v(A))^{(n-1)/n}.$$

But the latter relation actually yields the desired isoperimetric inequality. Indeed, this relation directly shows that, for a given value $s(A)$, the volume of A is bounded from above:

$$v(A) \leq (s(A)/\beta)^{n/(n-1)}.$$

Furthermore, it can easily be checked by a straightforward calculation that if A is a ball in $\mathbf{R}^n$, then we have the equality

$$s(A) = \beta(v(A))^{(n-1)/n}.$$

Thus, it remains to demonstrate that this equality is true only in the case when A is a ball. But the required result immediately follows from the next auxiliary statement.

Lemma 2. *Let A and B be any two compact convex bodies in the space $\mathbf{R}^n$ and let the relation*

$$s_B(A) = n(v(A))^{(n-1)/n}(v(B))^{1/n}$$

be satisfied. Then A and B are homothetic. In particular, if B is a ball, then A is a ball, too.

Proof. In virtue of the definition, we have

$$s_B(A) = lim_{r\to 0+}\frac{v(A+rB)-v(A)}{r}.$$

On the other hand, applying the formula of mixed volumes, we may write

$$v(A+rB) = \sum_{0\le k\le n}\frac{n!}{(n-k)!k!}r^k v_{n-k,k}(A,B).$$

Consequently,

$$v(A+rB) = v(A) + nrv(B,A,A,...,A) + o(r).$$

From this relation we get

$$s_B(A) = lim_{r\to 0+}\frac{v(A+rB)-v(A)}{r} = nv(B,A,A,...,A).$$

Thus, the relation

$$s_B(A) = n(v(A))^{(n-1)/n}(v(B))^{1/n}$$

is equivalent to the relation

$$(v(B,A,A,...,A))^n = (v(A))^{n-1}v(B).$$

In other words, we come to the case when the Minkowski inequality

$$(v(B,A,A,...,A))^n \ge (v(A))^{n-1}v(B)$$

is reduced to the equality. But we know from Section 2 that, in this case, the bodies A and B must be homothetic. This completes the proof of the lemma.

Therefore, we can formulate the following classical isoperimetric theorem (the proof of which has been presented above).

Theorem 2. *Among all compact convex bodies in the space* $\mathbf{R}^n$ $(n \geq 2)$ *with a given surface area, a ball (and only it) has a greatest volume.*

Note that Theorem 2 admits a number of generalizations to a much wider class of compact subsets of $\mathbf{R}^n$. For further extensions of this theorem, see e.g. [18], [47], [85], [178].

EXERCISES

1. Suppose that some strictly positive reals

$$a_1, a_2, \ldots, a_k \qquad (k \geq 3)$$

are given such that, for each natural index $i \in [1, k]$, the inequality

$$2a_i < \sum_{1 \leq j \leq k} a_j$$

is satisfied. Show that there are Jordan k-polygons in $\mathbf{R}^2$ whose sides have lengths respectively $a_1, a_2, \ldots, a_k$. Show also that, among such polygons, a polygon inscribed in a circle has a greatest area (in particular, this polygon is necessarily convex).

2. Let p be a fixed strictly positive real number. Consider the family of all those closed Jordan curves in $\mathbf{R}^2$ whose lengths are equal to p. Let L be any such curve. Prove that the compact subset of $\mathbf{R}^2$, which is homeomorphic to a circle and whose boundary coincides with L, is measurable in the Jordan sense. So we can speak of the area (in the Jordan sense) determined by this compact set. Demonstrate also that any circle whose boundary belongs to the above-mentioned family has a greatest area. This property is characteristic only for circles.

3. Let A be an arbitrary compact convex body in the Euclidean space $\mathbf{R}^n$. Prove that the usual surface area of this body is equal to the infimum of all surface areas of convex polyhedra containing A.

4. Give a proof of Lemma 1 from this section.

5. Consider the family F of all compact convex bodies in $\mathbf{R}^n$ whose diameters are equal to 1. Let us put

$$k = sup_{T \in F} v(T).$$

Show, by applying the Blaschke selection theorem, that there exists a body $T_0 \in F$ for which we have

$$v(T_0) = k,$$

and any T_0 with this property is a ball of diameter 1. Generalize this statement to the case of the family F' of all compact subsets of $\mathbf{R}^n$ whose diameters are equal to 1. We recall that, in the latter case, the volume of a compact subset of $\mathbf{R}^n$ is identified with its n-dimensional Lebesgue measure.

In connection with the result of Exercise 5, let us mention that this result and many others of a similar type are discussed in monograph [85] (see also the extensive bibliography in [85]).

6. Let X and Y be any two compact convex bodies of $\mathbf{R}^n$. Suppose that, for each n-dimensional simplex

$$[x_0, x_1, ..., x_n] \subset X,$$

there exists at least one translation g of $\mathbf{R}^n$ (certainly, depending on this simplex) for which

$$g([x_0, x_1, ..., x_n]) \subset Y.$$

Demonstrate that, in this case, there exists a translation f of $\mathbf{R}^n$ satisfying the inclusion

$$f(X) \subset Y.$$

7. Verify that any subset of the space $\mathbf{R}^n$ whose diameter is less than or equal to 1 can be covered by some closed ball whose radius is

$$r = (n/(2n+2))^{1/2}.$$

Check also that there are subsets of $\mathbf{R}^n$ with diameter 1 which cannot be covered by a ball of radius strictly smaller than r (for this purpose, consider a regular n-dimensional simplex in $\mathbf{R}^n$ all edges of which have length 1).

8. In the space $\mathbf{R}^n$, consider the family of all n-dimensional rectangular parallelepipeds whose surface area is equal to 1. Show that, among such parallelepipeds, a cube (and only it) has a greatest volume.

9. Let X be a compact convex body in $\mathbf{R}^n$, where $n \geq 2$, and let x be a point lying in the interior of X. For any hyperplane G in $\mathbf{R}^n$ passing

through x, denote by $X_1(G)$ and $X_2(G)$ the two subbodies of X determined by G and put

$$q(G) = \frac{v(X_1(G))}{v(X_2(G))}.$$

Let a hyperplane G_0 be chosen in such a way that $x \in G_0$ and

$$q(G_0) = sup_{G, x \in G}\ q(G).$$

Demonstrate that x coincides with the geometric centre of mass for the $(n-1)$-dimensional body $G_0 \cap X$.

10. Let $k \geq 4$ be a natural number and let T be a convex k-polygon in $\mathbf{R}^2$ with given angles. Denote by $s = s(T)$ the area of T and by $p = p(T)$ the perimeter of T. Applying the induction on k, show that the expression s/p^2 has a greatest value if and only if there exists a circle inscribed in T. Infer from this result the statement of Theorem 1 for $n = 2$.

11. Deduce directly from Theorem 1 the following weak version of Theorem 2: among all compact convex bodies in $\mathbf{R}^n$ with a given surface area, a ball has a greatest volume.

However, this way does not lead to the sharp assertion stating that a ball is a unique (up to motions) body in the above-mentioned family, at which volume attains its greatest value.

5. Some infinite-dimensional vector spaces

The next section of our book will be devoted to various kinds of Borel probability measures given on infinite-dimensional topological vector spaces (in particular, on infinite-dimensional Banach spaces). Among them some widely known classical spaces consisting of sequences of real numbers can be met rather frequently. Therefore, it is reasonable to recall in this section a number of typical properties of such spaces.

As usual, we denote by $\mathbf{N}$ the set of all natural numbers, i.e.

$$\mathbf{N} = \{0, 1, ..., n, ...\}.$$

The first space which is of interest to us is the product space $\mathbf{R}^{\mathbf{N}}$. It is also denoted by the symbol $\mathbf{R}^{\infty}$. Obviously, this is the largest space of sequences of real numbers. Note that $\mathbf{R}^{\mathbf{N}}$ can be regarded as a topological vector space with respect to the standard algebraic operations and with respect to the Tychonoff product topology. Also, it is not hard to see that $\mathbf{R}^{\mathbf{N}}$ is metrizable. Indeed, let us recall the following general theorem concerning the metrizability of topological vector spaces (cf., e.g., [175]).

Theorem 1. *Let E be a (Hausdorff) topological vector space. Then these two assertions are equivalent:*

1) E is metrizable;

2) the neutral element of E possesses a countable local base (i.e. has a countable fundamental system of neighbourhoods).

It immediately can be observed that the neutral element of $\mathbf{R}^{\mathbf{N}}$ has a countable local base, so $\mathbf{R}^{\mathbf{N}}$ is metrizable. Another, much more straightforward way to see the metrizability of this space is to utilize another well-known theorem from general topology, stating that a topological product of any countable family of metrizable spaces is metrizable, too (see, for example, [107]). Further, according to a result from the theory of uniform spaces, any product of complete uniform spaces is a complete uniform space (see again [107]). Thus $\mathbf{R}^{\mathbf{N}}$ is complete, as well. In addition, $\mathbf{R}^{\mathbf{N}}$ is separable, as a product of a countable family of separable spaces.

Let us also recall the following important

Definition 1. A topological vector space E is said to be a Frechet space if E is complete, metrizable and locally convex.

We thus see that $\mathbf{R}^{\mathbf{N}}$ is a separable Frechet space. On the other hand, it is not hard to check that $\mathbf{R}^{\mathbf{N}}$ does not possess any norm compatible with its topology. For this purpose, the definition of a bounded subset of a topological vector space E is helpful (see, e.g., [29] or [124]). Namely, a set $X \subset E$ is called bounded if, for each neighbourhood U of the neutral element of E, there exists a real $r > 0$ such that

$$(\forall t > r)(X \subset tU).$$

The following old result of Kolmogorov is true (cf. [124] or [175]).

Theorem 2. *For a (Hausdorff) topological vector space E, these two assertions are equivalent:*

1) there exists a norm on E compatible with the topology of E;

2) there exists at least one bounded convex neighbourhood of the neutral element of E.

Now, we easily observe that any neighbourhood of the neutral element of the space $\mathbf{R}^{\mathbf{N}}$ is not bounded. This fact implies at once that the structure of $\mathbf{R}^{\mathbf{N}}$ cannot be extended to the structure of a normed vector space.

The next notion (namely, the notion of a Polish topological space) is very important and plays a significant role in our further considerations.

Definition 2. A topological space E is called Polish if E is homeomorphic to some complete separable metric space.

In other words, E is Polish if its topological structure can be enriched to the structure of a complete separable metric space. It must be noted that Polish spaces are especially important in modern topological measure theory, because Borel measures defined on such spaces possess rather nice properties (see, e.g., [31] or Section 6 of this book).

Clearly, the space $\mathbf{R}^{\mathbf{N}}$ is Polish.

Of course, there are many complete metrics ρ compatible with the topology of the space $\mathbf{R}^{\mathbf{N}}$. A concrete example of such a metric ρ is as follows:

$$\rho(x, y) = \sum_{n \in \mathbf{N}} (1/2)^n \frac{|x_n - y_n|}{1 + |x_n - y_n|}$$

where $x = (x_n)_{n \in \mathbf{N}}$ and $y = (y_n)_{n \in \mathbf{N}}$ are any two elements of this space. The reader can easily check that ρ is a complete metric.

Let us describe the structure of the vector space of all linear continuous functionals on $\mathbf{R}^{\mathbf{N}}$. For each natural number n, denote by

$$pr_n : \mathbf{R}^{\mathbf{N}} \to \mathbf{R}$$

the canonical projection defined by the formula

$$pr_n(x) = x_n \qquad (x = (x_k)_{k \in \mathbf{N}} \in \mathbf{R}^{\mathbf{N}}).$$

Obviously, the family $(pr_n)_{n \in \mathbf{N}}$ consists of some linear functionals given on $\mathbf{R}^{\mathbf{N}}$. Let us denote by G the vector space generated by this family. Then $(\mathbf{R}^{\mathbf{N}}, G)$ is a pair of two vector spaces which are in a duality; more precisely, they are in duality with respect to the canonical bilinear functional

$$(x, g) \to g(x) \qquad (x \in \mathbf{R}^{\mathbf{N}},\ g \in G).$$

It is not hard to check that the topology of $\mathbf{R}^{\mathbf{N}}$ coincides with the weakest topology with respect to which all functionals from G become continuous. In other words, the topology of $\mathbf{R}^{\mathbf{N}}$ is identical with the weak topology $\sigma(\mathbf{R}^{\mathbf{N}}, G)$. Now, let us recall that the following general fact is known in the theory of topological vector spaces (cf., e.g., [29]).

Theorem 3. *Let E and F be any two vector spaces in duality with respect to some bilinear functional*

$$\phi : E \times F \to \mathbf{R}.$$

Let E be equipped with the weak topology $\sigma(E, F)$. Then the conjugate space E^* coincides with F (more precisely, E^* can be identified with F).

Applying this result in our particular case, we infer that the conjugate of the space $\mathbf{R}^{\mathbf{N}}$ coincides with the space G introduced above. In addition, we see that each continuous linear functional defined on $\mathbf{R}^{\mathbf{N}}$ depends only on finitely many coordinates. This fact is of interest and found numerous applications in probability theory and infinite-dimensional analysis.

Let us turn our attention to some vector subspaces of $\mathbf{R}^{\mathbf{N}}$. In the sequel, these subspaces will be equipped with certain topologies compatible with their vector structure but, in general, distinct from the topology induced by the topology of $\mathbf{R}^{\mathbf{N}}$.

Definition 3. We say that a sequence of real numbers $(x_n)_{n\in \mathbf{N}}$ has a finite support (or, simply, is finite) if it has only finitely many terms distinct from zero, i.e.

$$card(\{n \in \mathbf{N} \ : \ x_n \neq 0\}) < card(\mathbf{N}).$$

Evidently, the family of all finite sequences forms a vector subspace of $\mathbf{R}^{\mathbf{N}}$. This subspace is usually denoted by $\mathbf{R}^{(\mathbf{N})}$. It is clear that

$$\mathbf{R}^{(\mathbf{N})} = \cup_{n\in \mathbf{N}} \mathbf{R}^n$$

where

$$(\mathbf{R}^0, \mathbf{R}^1, ..., \mathbf{R}^n, ...)$$

is an increasing family of finite-dimensional subspaces of $\mathbf{R}^{\mathbf{N}}$ defined by the formula

$$x \in \mathbf{R}^n \Leftrightarrow (\forall k \geq n)(x_k = 0)$$

for any $n \in \mathbf{N}$. Further, it can immediately be checked that the space of finite sequences is everywhere dense in $\mathbf{R}^{\mathbf{N}}$. At the same time, the space of finite sequences is, in some sense, a minimal space in the family of all infinite-dimensional vector subspaces of $\mathbf{R}^{\mathbf{N}}$. More precisely, the algebraic dimension of $\mathbf{R}^{(\mathbf{N})}$ is equal to the first infinite cardinal $card(\mathbf{N})$, hence each vector subspace of $\mathbf{R}^{(\mathbf{N})}$ is either finite-dimensional or isomorphic to $\mathbf{R}^{(\mathbf{N})}$. We may introduce a canonical norm in $\mathbf{R}^{(\mathbf{N})}$ by the formula

$$||x|| = \sum_{n\in \mathbf{N}} |x_n| \qquad (x \in \mathbf{R}^{(\mathbf{N})}).$$

This definition of norm is correct because, in the sum written above, only finitely many members differ from zero. Nevertheless, it is not possible to introduce a norm in this space, with respect to which it becomes a Banach space. To see this, it suffices to point out again that $\mathbf{R}^{(\mathbf{N})}$ can be represented in the form of the union of a countable family of finite-dimensional vector subspaces. Each of these subspaces is nowhere dense in $\mathbf{R}^{(\mathbf{N})}$, so the classical Baire theorem on category fails to be true for $\mathbf{R}^{(\mathbf{N})}$. This implies the nonexistence of any complete norm for $\mathbf{R}^{(\mathbf{N})}$.

The second infinite-dimensional vector subspace of $\mathbf{R}^{\mathbf{N}}$ is much more interesting and important. It was introduced by Banach - one of the founders of modern functional analysis - who began his deep investigations with extensive study of properties of various vector subspaces of $\mathbf{R}^{\mathbf{N}}$ (see, for instance, his classical monograph [10]).

Definition 4. We shall say that a sequence $(x_n)_{n \in \mathbf{N}}$ from $\mathbf{R}^{\mathbf{N}}$ belongs to the space c_0 if there exists a limit $lim_{n \to \infty} x_n$ and, in addition, this limit equals zero.

For any sequence $x = (x_n)_{n \in \mathbf{N}}$ belonging to c_0, we put

$$||x||_\infty = sup_{n \in \mathbf{N}} |x_n|.$$

A direct verification shows that the functional $||\cdot||_\infty$ defined in this manner is a norm on c_0. Moreover, it is not difficult to check that $(c_0, ||\cdot||_\infty)$ has rather nice properties. Namely, we have the following statement.

Theorem 4. *The pair $(c_0, ||\cdot||_\infty)$ is a separable Banach space (in particular, a Polish space).*

The norm $||\cdot||_\infty$ is usually called the norm of uniform convergence. It is obvious that this norm can be canonically extended on the class of all bounded sequences of real numbers (which will be considered below).

Let us describe the structure of the conjugate space $(c_0)^*$. For this purpose, fix an arbitrary sequence $y = (y_n)_{n \in \mathbf{N}}$ from $\mathbf{R}^{\mathbf{N}}$ such that

$$||y||_1 = \sum_{n \in \mathbf{N}} |y_n| < +\infty$$

and, for each element $x \in c_0$, define

$$u_y(x) = \sum_{n \in \mathbf{N}} y_n x_n.$$

This definition is correct since the series on the right-hand side of the preceding equality converges absolutely. In this manner, we get a functional

$$u_y : c_0 \to \mathbf{R}$$

which is linear and continuous. Conversely, let u be an arbitrary continuous linear functional given on the space c_0. For each natural index n, let us denote by e_n the sequence whose all members are equal to zero, except the n-th member whose value equals 1. Further, we put

$$y_n = u(e_n) \qquad (n \in \mathbf{N}).$$

Taking into account the continuity of u, it is not hard to check that

$$\sum_{n \in \mathbf{N}} |y_n| < +\infty.$$

This also implies that $u = u_y$ where $y = (y_n)_{n \in \mathbf{N}}$. In other words, a mapping ϕ defined by the formula

$$\phi(y) = u_y$$

establishes a canonical one-to-one correspondence between the family of all absolutely summable sequences of real numbers and the family of all continuous linear functionals on c_0. Therefore, we may identify (in some sense) the space $(c_0)^*$ with the set of all absolutely summable sequences of real numbers. Note that this result follows also from the well-known Riesz theorem establishing a canonical one-to-one correspondence between signed Radon measures on a compact topological space T and linear continuous functionals on $C(T)$ (as usual, the symbol $C(T)$ denotes the Banach space of all continuous real-valued functions on T).

Definition 5. Let p be a strictly positive real number. We denote by l_p the set of all those sequences $(x_n)_{n \in \mathbf{N}}$ from the space $\mathbf{R}^{\mathbf{N}}$, for which

$$\sum_{n \in \mathbf{N}} |x_n|^p < +\infty.$$

For any sequence $x = (x_n)_{n \in \mathbf{N}}$ from l_p, we put

$$||x||_p = (\sum_{n \in \mathbf{N}} |x_n|^p)^{1/p}.$$

It can easily be shown that

$$||x + y||_p \leq ||x||_p + ||y||_p$$

for $p \geq 1$ and for any two elements x and y from l_p. The above-mentioned inequality follows directly from the elementary fact that the function

$$f(t) = (1 - t^{1/p})^p \qquad (0 \leq t \leq 1)$$

is convex whenever $p \geq 1$. Thus, if $p \geq 1$, the set l_p becomes a vector space and $|| \cdot ||_p$ turns out to be a norm on l_p. Moreover, we have the following classical Riesz-Fisher theorem (see, e.g., [59] or [124]).

Theorem 5. *If $p \geq 1$, then $(l_p, || \cdot ||_p)$ is a separable Banach space (in particular, a Polish space).*

According to this result, l_1 is a separable Banach space with respect to the norm $|| \cdot ||_1$. Actually, l_1 coincides with the family of all absolutely summable sequences of real numbers. As pointed out above, we may identify l_1 with the conjugate space of c_0, i.e. we may write

$$(c_0)^* = l_1.$$

Note, in addition, that the space of all finite sequences $\mathbf{R}^{(\mathbf{N})}$ is contained in each of the spaces l_p as an everywhere dense subset of l_p.

Now, let us describe the structure of the space $(l_1)^*$. For this purpose, fix any bounded sequence $y = (y_n)_{n\in\mathbf{N}}$ of real numbers. Then, for each element $x = (x_n)_{n\in\mathbf{N}}$ from l_1, define

$$u_y(x) = \sum_{n\in\mathbf{N}} x_n y_n.$$

This definition of $u_y(x)$ is correct since the series on the right-hand side of the preceding equality is absolutely convergent. Furthermore, it can easily be seen that the functional

$$u_y \ : \ l_1 \to \mathbf{R}$$

is linear and continuous on l_1. Also, a simple argument shows that the norm of u_y is

$$||u_y|| = sup_{n\in\mathbf{N}}|y_n|.$$

Conversely, suppose that u is an arbitrary linear continuous functional on l_1. Considering again, for each $n \in \mathbf{N}$, the element e_n of l_1 and putting

$$y_n = u(e_n) \qquad (n \in \mathbf{N}),$$

we infer that the sequence $y = (y_n)_{n\in\mathbf{N}}$ is bounded and hence the equality $u = u_y$ holds true. In other words, a mapping ψ defined by the formula

$$\psi(y) = u_y$$

establishes a canonical one-to-one correspondence between the family of all bounded sequences of real numbers and the family of all linear continuous functionals on l_1.

Definition 6. The set of all bounded sequences from the space $\mathbf{R}^{\mathbf{N}}$ is denoted by l_∞. It is a vector subspace of $\mathbf{R}^{\mathbf{N}}$ and is equipped with the norm of uniform convergence:

$$||y||_\infty = sup_{n\in\mathbf{N}}|y_n|$$

for any $y \in l_\infty$.

We have the following simple statement.

Theorem 6. *The pair $(l_\infty, ||\cdot||)$ is a nonseparable Banach space.*

Indeed, it is clear that l_∞ is a Banach space. In order to demonstrate that this space is nonseparable as well, it suffices to check the nonseparability of the closed unit ball

$$B = \{y \in l_\infty \ : \ ||y||_\infty \leq 1\}.$$

For each subset P of $\mathbf{N}$, consider the characteristic function (i.e. indicator) of P and denote it by y_P. Clearly, y_P is a sequence whose terms belong to $\{0,1\}$. Hence $y_P \in B$. Further, the relations

$$P \subset \mathbf{N}, \quad Q \subset \mathbf{N}, \quad P \neq Q$$

imply the equality

$$||y_P - y_Q||_\infty = 1.$$

Now, since the family $(y_P)_{P \subset \mathbf{N}}$ is of cardinality continuum, we conclude that the ball B is nonseparable. Hence l_∞ is nonseparable, too. We leave to the reader, as a useful exercise, to find all extreme points of B. In this connection, it must also be noted that the closed unit ball in the space c_0 does not have extreme points.

As shown earlier, the space l_∞ can be identified with the conjugate of l_1. Thus, we come to the equalities

$$(l_1)^* = l_\infty, \qquad (c_0)^{**} = l_\infty.$$

The investigation of the conjugate of l_∞ is much more difficult than in the previous cases. The difficulty is implied by the circumstance that, starting with the Axiom of Choice (or the Zorn lemma), some continuous linear functionals on l_∞ with rather strange properties can be defined. In fact, the conjugate of l_∞ is closely connected with the family of all so-called finitely additive signed measures given on $\mathbf{N}$. We want to finish this section with a construction of a functional of the above-mentioned type. Such functionals were first constructed by Banach and his argument was one of the earliest examples of utilizing essentially transfinite methods in actual problems of functional analysis (cf., e.g., the historical comments given in [29]).

We need a Hahn-Banach type theorem. Let E be a vector space and let

$$p \ : \ E \to \mathbf{R}$$

be a functional satisfying the conditions:

1) $p(x+y) \leq p(x) + p(y)$ for all x and y from E;

2) $p(tx) = tp(x)$ for each $x \in E$ and for any $t \geq 0$.

Suppose also that a vector space $F \subset E$ and a linear functional

$$h_0 : F \to \mathbf{R}$$

are given such that

$$h_0(x) \leq p(x) \qquad (x \in F).$$

Then, according to the analytic form of the Hahn-Banach theorem (see Theorem 2 from Section 1), there exists a linear functional

$$h : E \to \mathbf{R}$$

such that

$$h(x) \leq p(x) \qquad (x \in E).$$

We are going to apply this result to the space $E = l_\infty$. For each natural number n, define a linear functional

$$g_n : l_\infty \to \mathbf{R}$$

by the formula

$$g_n(x) = \frac{x_0 + ... + x_n}{n+1} \qquad (x \in l_\infty).$$

In other words, $g_n(x)$ coincides with the arithmetic mean of the first $n+1$ terms of $x \in l_\infty$. Further, let us put

$$p(x) = limsup_{n \to \infty} g_n(x) \qquad (x \in l_\infty).$$

Note that this definition of $p(x)$ is correct since

$$|g_n(x)| \leq ||x||_\infty \qquad (n \in \mathbf{N},\ x \in l_\infty),$$

which also implies that

$$|p(x)| \leq ||x||_\infty \qquad (x \in l_\infty)$$

and, in particular, $p(x)$ is always finite. Let us denote by F the set of all those $x \in l_\infty$ for which there exists a limit

$$lim_{n \to \infty} g_n(x) = h_0(x).$$

Clearly, F is some vector subspace of l_∞ and h_0 is some linear functional on F. Furthermore, in virtue of the definition of h_0, we have

$$h_0(x) = p(x) \qquad (x \in F).$$

Now, we must check that p satisfies conditions 1) and 2). Indeed, it is well known that if $(a_n)_{n\in\mathbf{N}}$ and $(b_n)_{n\in\mathbf{N}}$ are two arbitrary bounded sequences of real numbers, then

$$limsup_{n\to\infty}(a_n + b_n) \leq limsup_{n\to\infty}a_n + limsup_{n\to\infty}b_n.$$

This immediately implies that, for any x and y from l_∞, we have

$$p(x+y) = limsup_{n\to\infty}(g_n(x) + g_n(y)) \leq$$

$$limsup_{n\to\infty}g_n(x) + limsup_{n\to\infty}g_n(y) = p(x) + p(y).$$

Besides, it is evident that if $x \in l_\infty$ and $t \geq 0$, then

$$p(tx) = tp(x).$$

Consequently, there exists a linear functional

$$h : l_\infty \to \mathbf{R}$$

extending h_0 and satisfying the relation

$$h(x) \leq p(x) \qquad (x \in l_\infty).$$

The functional h is continuous since

$$|h(x)| \leq ||x||_\infty \qquad (x \in l_\infty).$$

Furthermore, an easy argument shows that, for any $x = (x_n)_{n\in\mathbf{N}}$ from l_∞, the inequality

$$p(x) \leq limsup_{n\to\infty}x_n$$

holds true. This implies that

$$h(x) \leq limsup_{n\to\infty}x_n.$$

Taking into account the fact that

$$liminf_{n\to\infty}x_n = -limsup_{n\to\infty}(-x_n)$$

and utilizing the linearity of h, we finally get

$$liminf_{n\to\infty}x_n \le h(x) \le limsup_{n\to\infty}x_n.$$

In particular, we can conclude that if a sequence $x = (x_n)_{n\in\mathbf{N}}$ is convergent, then

$$h(x) = lim_{n\to\infty}x_n.$$

In other words, the linear functional h is an extension of the usual linear functional lim defined on the vector space of all convergent sequences of real numbers. Accordingly, h is often called a generalized Banach limit.

Let us point out one interesting property of h. For this purpose, consider in l_∞ the so-called shift operator

$$s : l_\infty \to l_\infty$$

defined by the formula

$$(s(x))_n = x_{n+1} \qquad (x \in l_\infty,\ n \in \mathbf{N}).$$

For any x from l_∞, we may write

$$h(x - s(x)) \le p(x - s(x)).$$

On the other hand, we also have

$$p(x - s(x)) = limsup_{n\to\infty}\frac{x_0 - x_{n+1}}{n+1} = 0.$$

Consequently,

$$h(x - s(x)) \le 0.$$

Replacing, in the inequality above, x by $-x$ and applying the linearity of h, we obtain

$$h(x - s(x)) \ge 0.$$

Summarizing all the preceding results, we come to the equality

$$h(x - s(x)) = 0 \qquad (x \in l_\infty)$$

or, equivalently,

$$h(x) = h(s(x)) \qquad (x \in l_\infty),$$

which means that the functional h is invariant with respect to the shift operator.

As mentioned earlier, the presented method of constructing generalized limits on the space l_∞ is due to Banach (cf. [10]). This method can be applied in many other situations. In particular, by using a much more complicated argument, Banach was able to demonstrate the existence of an universal finitely additive extension of the Lebesgue measure on $\mathbf{R}^2$, which possesses the invariance property with respect to the group of all isometric transformations of $\mathbf{R}^2$ (for more detailed information concerning this topic, see e.g. [29], [113] and [199]). Note that, according to the classical result of Hausdorff, for all natural numbers $n \geq 3$, there are no universal finitely additive measures on $\mathbf{R}^n$ invariant under the group of all isometric transformations of $\mathbf{R}^n$ (see [90]; cf. also [113] and [199]).

EXERCISES

1. Let E be an arbitrary infinite-dimensional Frechet space. Prove that the algebraic dimension of E is greater than or equal to the cardinality continuum.

2. Let I be any set of cardinality less than or equal to the cardinality continuum. Consider the topological vector space $\mathbf{R}^I$. Demonstrate that this space is separable, i.e. there exists a countable everywhere dense subset of $\mathbf{R}^I$ (however, if I is uncountable, then the space $\mathbf{R}^I$ does not have a countable base and, moreover, no point of this space possesses a countable local base).

3. Let I be an arbitrary set. Consider again the topological vector space $\mathbf{R}^I$. Show that this space satisfies the so-called Suslin condition, i.e. any disjoint family of nonempty open subsets of $\mathbf{R}^I$ is at most countable. For this purpose, identify the above-mentioned space with some everywhere dense subset of the product space S^I, where S is a one-dimensional torus (regarded as a compact commutative topological group) and apply to the compact topological group S^I the theorem of the existence of a Haar probability measure (see, e.g., [86]).

4. Describe the Banach space of all linear continuous mappings acting from c_0 into itself.

5. Describe the Banach space of all linear continuous mappings acting from l_1 into itself.

6. Demonstrate that there exists, in c_0, a closed hyperplane H having the property that there is no affine hyperplane in c_0 parallel to H and support for the closed unit ball of c_0. (cf. Theorem 6 from Section 1).

7. Show that, for any locally convex topological vector space E, the following two assertions are equivalent:

a) the space E is metrizable;

b) the topology of E can be determined by some countable family of seminorms.

8. Demonstrate that every locally convex topological vector space is isomorphic to some subspace of a product of Banach spaces.

9. As shown in this section, $(l_1)^* = l_\infty$. Check that $(l_\infty)^* \neq l_1$ and conclude that l_1 is not a reflexive Banach space.

10. Let $p > 1$. Prove that $(l_p)^* = l_q$ where q is defined by the equality

$$1/p + 1/q = 1.$$

Deduce from this fact that $(l_p)^{**} = l_p$, i.e. the Banach space l_p is reflexive.

11. Demonstrate that there exists at least one finitely additive measure μ defined on the class of all subsets of the plane $\mathbf{R}^2$, such that:

a) μ extends the standard two-dimensional Lebesgue measure λ_2 on $\mathbf{R}^2$;

b) μ is invariant with respect to the transformation group generated by the family of all central symmetries of $\mathbf{R}^2$.

In this connection, let us note once more that, according to the classical result of Banach, μ can be chosen to be invariant even with respect to the group of all isometric transformations of $\mathbf{R}^2$, which is essentially wider than the group of b). However, the group of b) is large, too, because it contains all translations of $\mathbf{R}^2$.

12. Let $0 < p < 1$. Consider in the topological vector space l_p a countable family $(e_i)_{i \geq 1}$ of points, defined as follows: $e_i(n) = n^{p-1}$ if $n = i$, and $e_i(n) = 0$ if $n \neq i$. Show that the set

$$X = \{0, e_1, e_2, ..., e_i, ...\} \subset l_p$$

is compact, but the closed convex hull of X is not bounded in l_p. In particular, this closed convex hull is not compact.

13. Let $(x_n)_{n \in \mathbf{N}}$ be a sequence of elements from l_1. Prove that the following two assertions are equivalent:

a) $(x_n)_{n \in \mathbf{N}}$ is convergent in the norm of l_1;

b) $(x_n)_{n \in \mathbf{N}}$ is weakly convergent.

This means that, for the space l_1, the two kinds of convergence of sequences - strong and weak - turn out to be the same. On the other hand, show that the strong (i.e. determined by the norm) topology of l_1 differs from the weak topology $\sigma(l_1, (l_1)^*)$, i.e. the equivalence of a) and b) does not hold for generalized sequences of elements from l_1.

14. Let Q be a compact topological space. Denote by $C(Q)$ the Banach space of all continuous real-valued functions defined on Q (as usual, we equip $C(Q)$ with the norm of uniform convergence). Demonstrate that the next two statements are equivalent:

a) $C(Q)$ is separable;

b) Q has a countable base (i.e. Q is metrizable).

15. Let E be a Hilbert space (over $\mathbf{R}$). Show that the following two assertions are equivalent:

a) E is infinite-dimensional;

b) there exist a subset X of E and an isometric embedding

$$\Psi : X \to E,$$

such that Ψ cannot be extended to an injection of E into itself.

16. Let E be an infinite-dimensional Hilbert space (over $\mathbf{R}$). Demonstrate that there exists an injective continuous mapping

$$f : [0,1] \to E$$

having the property that, for any four points t_1, t_2, t_3, t_4 from $[0,1]$ satisfying the relation

$$t_1 < t_2 \leq t_3 < t_4,$$

the linear segments $[f(t_1), f(t_2)]$ and $[f(t_3), f(t_4)]$ are perpendicular in E.

Note that the set $f([0,1])$ is a curve in E homeomorphic to $[0,1]$. It is usually called the Wiener curve.

17. Let $(E, ||\cdot||)$ be a separable normed vector space and let

$$K = \{g \in E^* : ||g|| \leq 1\}$$

denote the closed unit ball in E^*. Equip K with the topology induced by the weak topology $\sigma(E^*, E)$. Prove that:

a) K is a compact metrizable space;

b) there exists a continuous mapping from $[0,1]$ onto K.

Deduce from these facts the classical Banach-Mazur theorem stating the universality of $C[0,1]$ in the class of all separable metric spaces; in other words, show that, for any separable metric space X, there exists a subset of $C[0,1]$ isometric to X.

6. Probability measures and random elements

One of the central topics in our book can be characterized as follows: we discuss appropriate analogues of the Anderson inequality (see Section 2) and demonstrate applications of these analogues in probability theory, mathematical statistics and infinite-dimensional analysis. In other words, we wish to consider several problems and questions which are closely connected with the Anderson inequality and are of interest from the probabilistic viewpoint. This consideration essentially relies on the techniques from contemporary measure theory and probability theory. Therefore, it is reasonable to recall here some preliminary material concerning probability measures and random elements (for more detailed information, we refer the reader to [14], [31], [57], [71], [74], [80], [86], [155], [161], [171], [180], [197]).

Let G be a set and let S be a σ-algebra of its subsets. The pair (G, S) is usually called a measurable space. Elements of S are called measurable (more precisely, S-measurable) subsets of G.

Let K be some class of subsets of G. We denote by $\sigma(K)$ the smallest (with respect to inclusion) σ-algebra of subsets of G, containing K. This σ-algebra is called the σ-algebra generated by the class K. Obviously, it coincides with the intersection of all those σ-algebras in G which contain K.

It is clear that G can be equipped with the structure of a measurable space in various ways: it suffices to consider the σ-algebras generated by different classes of subsets of G. Notice that the σ-algebras associated with certain topologies on G are of special interest to us.

Let F be a topological space. We denote by $B(F)$ the σ-algebra generated by the family of all open (equivalently, closed) subsets of F. As usual, $B(F)$ is called the Borel σ-algebra of F and the pair $(F, B(F))$ is called a Borel space. Elements of $B(F)$ are called Borel subsets of F.

Example 1. Let F be an arbitrary metric space and let K be the family of all open balls in F. Evidently, we have the inclusion

$$\sigma(K) \subset B(F).$$

In general, this inclusion is proper. But if the space F is separable, then $\sigma(K) = B(F)$. The same equality remains true when K is a countable family of balls which form a base of the topology of F.

Further, let G be a set, let (G', S') be a measurable space and let $(X_j)_{j\in J}$ be a family of mappings acting from G into G'. We denote by $\sigma((X_j)_{j\in J}, S')$ the σ-algebra in G generated by the class

$$\{X_j^{-1}(A) \; : \; j \in J, \; A \in S'\}.$$

This σ-algebra is usually called the σ-algebra generated by the family $(X_j)_{j\in J}$ (or by all mappings from this family). There is another, equivalent way to introduce this σ-algebra. We indicate it below.

Let (G, S) and $(G,' S')$ be two measurable spaces and let

$$X \; : G \to G'$$

be a mapping. We say that X is measurable (more precisely, (S, S')-measurable) if $X^{-1}(S') \subset S$. Obviously, the σ-algebra $\sigma((X_j)_{j\in J}, S')$ can be characterized as a smallest σ-algebra in G with respect to which all mappings X_j $(j \in J)$ become measurable.

The following particular situation is of extreme importance.

Let $(F, B(F))$ and $(F', B(F'))$ be any two Borel spaces. A measurable mapping from F into F' is called a Borel mapping. Clearly, every continuous mapping from F into F' is Borel. Consequently, if $C(F)$ denotes the family of all continuous real-valued functions defined on F, then the σ-algebra $\sigma(C(F), B(\mathbf{R}))$ is contained in $B(F)$. It turns out that, under certain natural assumptions on F, the Borel σ-algebra of F is generated by rather small subfamilies of $C(F)$ (cf. Exercise 1 to this section).

Let F be a topological vector space and let F^* denote its conjugate space. In our further considerations we assume, as a rule, that F^* separates the points of F. For any $H \subset F^*$, we shall denote by $B(F, H)$ the σ-algebra in F generated by all functionals from H. In other words, we put

$$B(F, H) = \sigma(H, B(\mathbf{R})).$$

Lemma 1. *Let $(F, ||\cdot||)$ be a separable Banach space. Then there exists a countable set $H \subset F^*$ for which we have*

$$B(F) = B(F, H).$$

Proof. Indeed, according to a consequence of the Hahn-Banach theorem (for separable normed vector spaces), there exists a countable set $H \subset F^*$ such that

$$||x|| = sup_{h \in H}|h(x)| \quad (x \in F).$$

This immediately implies that all the balls

$$U_r(y) = \{x \in F \ : \ ||x - y|| < r\} \qquad (y \in F, \ r > 0)$$

are $B(F, H)$-measurable. Applying again the separability of F, we get at once the required equality $B(F) = B(F, H)$.

The lemma just proved is a very particular case of the general result stating that any countable subset of F^* separating the points of F can be taken as H (see Exercise 1).

Let us return to the case when F is an arbitrary topological vector space. Fix a set $H \subset F^*$ and consider any finite family

$$H' = \{h_1, \ h_2, \ ..., \ h_m\} \subset H.$$

This family generates the σ-algebra $\sigma(H', B(\mathbf{R}))$ in the space F, whose elements are sets of the form

$$C_{h_1,h_2,...,h_m}(D) = \{x \in F \ : \ (h_1(x), h_2(x), ..., h_m(x)) \in D\}$$

where D ranges over $B(\mathbf{R}^m)$. These sets are usually called cylindrical ones. Clearly, the family

$$\{\sigma(H', B(\mathbf{R})) \ : \ H' \subset H \ and \ H' \ is \ finite\}$$

is directed with respect to the standard inclusion relation and the union of this family forms a certain algebra of sets , which is called the algebra of H-cylindrical sets in F. From now on, it will be denoted by the symbol $S(F, H)$. Accordingly, the σ-algebra generated by $S(F, H)$ is called the σ-algebra of H-cylindrical subsets of F and is denoted by the symbol $\sigma(S(F, H))$. In many cases below, H coincides with F^* and we simply speak of the cylindrical algebra (cylindrical σ-algebra) of a given space F. It is not hard to check that the following two relations hold true:

1) $S(F, H) = S(F, lin(H))$ where $lin(H)$ denotes the vector space generated by the family H;

2) $\sigma(S(F, H)) = B(F, H)$.

Relation 2) and Lemma 1 also imply the next auxiliary proposition which turns out to be useful in various situations.

Lemma 2. *Let F be a separable Banach space. Then*

$$B(F) = \sigma(S(F, F^*)).$$

More generally, for any countable set $H \subset F^*$ separating the points of F, we have

$$B(F) = \sigma(S(F, H)).$$

A result essentially extending Lemma 2 is presented in Exercise 1.

Let us consider several examples connected with the properties of algebras and σ-algebras of cylindrical sets in some classical spaces of sequences of real numbers.

Example 2. Let $F = \mathbf{R}^{\mathbf{N}}$ and assume that F is equipped with the standard product topology. Denote $H = (pr_n)_{n \in \mathbf{N}}$ where

$$pr_n \; : \; \mathbf{R}^{\mathbf{N}} \to \mathbf{R}$$

is the canonical projection corresponding to the n-th coordinate. The algebra $S(F, H)$ consists of all cylindrical sets which admit the representation

$$\{(x_n)_{n \in \mathbf{N}} \in F \; : \; (x_0, \ldots, x_m) \in D\}$$

where $m \in \mathbf{N}$ and D is a Borel subset of $\mathbf{R}^{m+1}$. Since the conjugate of F coincides with the linear hull of H, the algebra of all cylindrical sets in F is identical with the algebra of all H-cylindrical sets. Furthermore, since F is a separable Frechet space, we have

$$B(F) = B(F, H).$$

Note that the preceding equality follows immediately from one general result which is formulated in Exercise 1. But this equality can be established directly, by taking into account the fact that the family of all sets of the form

$$\{(x_n)_{n \in \mathbf{N}} \; : \; |x_0| < \varepsilon, \ldots, |x_m| < \varepsilon\},$$

where $m \in \mathbf{N}$ and $\varepsilon > 0$, is a fundamental system of neighbourhoods of zero in F. Further, if $G \subset F$, then the family of sets

$$(G \cap D)_{D \in B(F)}$$

is a σ-algebra of subsets of G. We shall denote it by the symbol $B_1(G)$. Actually, $B_1(G)$ is the σ-algebra in G induced by $B(F)$. It should be observed that

$$G \in B(F) \Rightarrow B_1(G) \subset B(F),$$

$$B_1(G) = \sigma((G \cap D)_{D \in S(F,H)}).$$

Example 3. Let us put again $F = \mathbf{R}^{\mathbf{N}}$ and consider the space $c_0 \subset F$. It is not hard to check the relation

$$c_0 = \cap_{k \in \mathbf{N}} \cup_{m \in \mathbf{N}} \{(x_n)_{n \in \mathbf{N}} \in \mathbf{R}^{\mathbf{N}} : |x_m| < \frac{1}{k+1}, |x_{m+1}| < \frac{1}{k+1}, ...\}.$$

This relation shows, in particular, that c_0 is a Borel subset of F. Moreover, as we know, c_0 is a separable Banach space with respect to the norm of uniform convergence. In addition, we recall that $(c_0)^* = l_1$.

We may consider the following four σ-algebras in c_0:

$$B_1(c_0), \quad B(c_0, H), \quad B(c_0, (c_0)^*), \quad B(c_0),$$

where $H = (pr_n)_{n \in \mathbf{N}}$. Clearly, $B_1(c_0) = B(c_0, H)$, which implies that $B(c_0, H) \subset B(F)$. We also have the inclusions

$$B(c_0, H) \subset B(c_0, (c_0)^*) \subset B(c_0).$$

Actually, it turns out that these four σ-algebras coincide. To see this, it suffices to note that any open ball in c_0 is $B(c_0, H)$-measurable and then to apply the separability of the space c_0.

Example 4. Put again $F = \mathbf{R}^{\mathbf{N}}$ and consider the space $l_\infty \subset F$. As we know, l_∞ is a nonseparable Banach space with respect to the norm of uniform convergence. Similarly to the previous example, we may introduce the following four σ-algebras:

$$B_1(l_\infty), \quad B(l_\infty, H), \quad B(l_\infty, (l_\infty)^*), \quad B(l_\infty),$$

where $H = (pr_n)_{n \in \mathbf{N}}$. The space l_∞ and each open ball in it are Borel subsets of F (the reader can check this fact without any difficulty). Also, it is clear that

$$B_1(l_\infty) = B(l_\infty, H).$$

Hence we have the inclusion

$$B(l_\infty, H) \subset B(F).$$

But here there is no further analogy with the space c_0, because all the inclusions

$$B(l_\infty, H) \subset B(l_\infty, (l_\infty)^*) \subset B(l_\infty)$$

turn out to be proper (see Exercise 2). We also wish to point out especially that

$$B(c_0) \subset B(l_\infty, H).$$

A significant difference between the structures of cylindrical and Borel σ-algebras of the space l_∞ implies the fact that the problem of constructing suitable Borel probability measures on l_∞ is rather delicate. This problem is important for probability theory and is closely connected with some interesting properties of sequences of random variables. We shall deal with these questions in our further considerations.

Now, let us turn our attention to measurable subsets of the space $\mathbf{R}^T$ where T is an arbitrary (possibly, uncountable) set of indices. As usual, elements of this space will be denoted by $x = (x_t)_{t \in T}$.

Example 5. Put $F = \mathbf{R}^T$, where T is uncountable, and equip F with the product topology. Denote also $H = (pr_t)_{t \in T}$ where the canonical linear functional pr_t $(t \in T)$ is defined by

$$pr_t(x) = x_t \qquad (x \in F).$$

Since

$$F^* = (\mathbf{R}^T)^* = lin(H),$$

we infer that

$$S(F, H) = S(F, F^*).$$

Elements of $S(F, H)$ are of the form

$$\{x \in F \ : \ (x_{t_1}, ..., x_{t_m}) \in D\}$$

where $m \in \mathbf{N}$, $(t_1, ..., t_m)$ is an injective finite subfamily of T and D is a Borel subset of $\mathbf{R}^m$. We thus see that each element of the canonical base of the topology of F is $B(F, H)$-measurable. Nevertheless, it is not hard to demonstrate that the σ-algebra $B(F, H)$ is properly contained in the Borel σ-algebra $B(F)$. This fact does not contradict the result of Exercise 1 of this section. Indeed, in accordance with Theorem 1 from Section 5, the assumption $card(T) > card(\mathbf{N})$ implies that the space $F = \mathbf{R}^T$ is not metrizable. We recall that this space is separable whenever $card(T)$ is less than or equal to the cardinality continuum (see Exercise 2 from Section 5).

Example 6. Put again $F = \mathbf{R}^T$ and let $L_\infty(T)$ denote the subspace of F consisting of all bounded functions acting from T into $\mathbf{R}$. This space may be regarded as a Banach space with respect to the norm of uniform convergence. Clearly, $L_\infty(T)$ is nonseparable for any infinite T. In addition, if T is uncountable, then $L_\infty(T)$ is not a $B(F, F^*)$-measurable subset of F. An analogous situation can be observed for the space $C(T)$ of all real-valued continuous functions on T, where T is some uncountable topological space. However, if T is a compact metric space, then $C(T)$ becomes a separable Banach space with respect to the norm of uniform convergence, and we have the equality

$$B(C(T), (pr_t)_{t \in Q}) = B(C(T))$$

where Q is an arbitrary countable everywhere dense subset of T (in this connection, cf. Lemma 1).

Now, let us recall some basic notions from classical measure theory.

Let G be a set and let S be some algebra of subsets of G. A functional

$$\mu : S \to [0, +\infty]$$

is called a premeasure (or finitely additive measure) on S if

$$\mu(D_1 \cup D_2) = \mu(D_1) + \mu(D_2)$$

for any two disjoint sets D_1 and D_2 from S. We say that a premeasure μ is a measure on S if

$$\mu(\cup\{D_n : n \in \mathbf{N}\}) = \sum_{n \in \mathbf{N}} \mu(D_n)$$

for any disjoint countable family $\{D_n : n \in \mathbf{N}\} \subset S$ whose union also belongs to S.

We say that a premeasure μ is finite if $\mu(G) < +\infty$.

We say that a premeasure μ is σ-finite if there exists at least one covering $\{D_n : n \in \mathbf{N}\}$ of G consisting of elements from S, such that

$$(\forall n \in \mathbf{N})(\mu(D_n) < +\infty).$$

A given premeasure (measure) μ is a probability premeasure (measure) if $\mu(G) = 1$. In our further considerations, we primarily deal with probability premeasures and probability measures.

Let S and S' be two algebras of subsets of G, such that $S \subset S'$. We say that a premeasure μ defined on S can be extended to a premeasure defined on S' if there exists at least one premeasure μ' with $dom(\mu') = S'$ such that

$$(\forall D \in S)(\mu'(D) = \mu(D)).$$

In this case, we also say that μ' is an extension of μ (and μ is a restriction of μ'). If S' coincides with the σ-algebra generated by S, i.e. $S' = \sigma(S)$, then the problem of extending a given premeasure on S to a measure on S' is solved by Carathéodory whose classical theorem states that a premeasure μ on S is extendable to a measure μ' on S' if and only if μ is a measure on S (see, e.g., [59], [86], [124], [155], [177]).

By using the Carathéodory theorem, the existence of a product measure can be established for any family of probability measures.

Example 7. Let us consider a family

$$((G_j, S_j, \mu_j))_{j \in J}$$

of spaces equipped with probability measures. This means that, for each index $j \in J$, the pair (G_j, S_j) is a measurable space and μ_j is a probability measure with $dom(\mu_j) = S_j$. A new probability measure space

$$(G, S, \mu) = (\prod_{j \in J} G_j, \prod_{j \in J} S_j, \prod_{j \in J} \mu_j)$$

can be canonically associated with this family. Here $G = \prod_{j \in J} G_j$ denotes the Cartesian product of the family of sets $(G_j)_{j \in J}$, the symbol $S = \prod_{j \in J} S_j$ denotes the smallest σ-algebra in G with respect to which all projections pr_j $(j \in J)$ are measurable, and

$$\mu = \prod_{j \in J} \mu_j$$

is the product measure on S obtained from the family $(\mu_j)_{j \in J}$ of given probability measures (actually, the product measure μ is uniquely determined by the condition that an analogue of the classical Fubini theorem must be fulfilled for μ and $(\mu_j)_{j \in J}$). Also, the reader can easily observe that S is very similar to cylindrical σ-algebras introduced earlier.

In particular, if $J = \mathbf{N}$ and

$$(G_j, S_j) = (\mathbf{R}, B(\mathbf{R}))$$

for all $j \in \mathbf{N}$, we come to the Borel probability product measure

$$\mu = \prod_{n \in \mathbf{N}} \mu_n$$

on the Polish topological vector space $\mathbf{R}^{\mathbf{N}}$.

The construction of a product measure is frequently applied in analysis and probability theory. We shall meet this construction many times in subsequent sections of the book.

A direct verification of the assumptions of the Carathéodory theorem mentioned above is not easy, in general, and some special cases of this theorem are more convenient for applications. Let us describe several such situations. For this purpose, consider a topological vector space F and fix a subset H of F^*. A premeasure μ defined on the algebra $S(F, H)$ is called a cylindrical (more precisely, H-cylindrical) premeasure if, for any finite family of functionals

$$\{h_1, h_2, ..., h_m\} \subset H,$$

the restriction of μ to the σ-algebra $B(F, \{h_1, h_2, ..., h_m\})$ turns out to be a measure on this σ-algebra.

Example 8. Let T be a set of indices and let $F = \mathbf{R}^T$. We take again

$$H = (pr_t)_{t \in T}.$$

If the set T is finite, then the above definition implies that any H-cylindrical premeasure on $B(F, H)$ is a measure. Moreover, one of the consequences of the famous Kolmogorov theorem (see, e.g., [121], [155], [171] or [180]) states that, for every T (not necessarily finite), each H-cylindrical premeasure admits an extension to a measure defined on the cylindrical σ-algebra $B(F, H)$, and the latter measure is unique. Recall now that if T is infinite and countable simultaneously, then $B(F, H)$ coincides with the Borel σ-algebra $B(F)$. In this case, we can conclude that every H-cylindrical premeasure admits a unique extension to a Borel measure on F. In other words, for a countable T, the notion of an H-cylindrical premeasure is reduced to the notion of a Borel measure. For uncountable sets T, such a reduction cannot be done.

We know that the product space $F = \mathbf{R}^{\mathbf{N}}$ may be treated as a separable Frechet space. So a natural question arises whether it is possible to generalize the fact described in Example 8 to the class of all separable Frechet spaces (and to other classes of topological vector spaces). In order to consider briefly this question, we need one useful result concerning Borel measures on Polish topological spaces.

Let F be a topological space and let μ be a Borel measure on F. We say that μ is a Radon measure if

$$\mu(D) = sup\{\mu(K) \ : \ K \subset D \ and \ K \ is \ compact\}$$

for all Borel sets $D \subset F$.

We say that a topological space F is Radon if each σ-finite Borel measure on F is Radon.

The following classical statement is due to Ulam.

Theorem 1. *If F is a Polish space, then F is Radon.*

Proof. Let μ be an arbitrary σ-finite Borel measure on F. We have to show that μ is a Radon measure. If μ is identically zero, then there is nothing to prove. Suppose now that $\mu \neq 0$ and let μ' denote a Borel probability measure on F equivalent to μ (i.e. the class of all μ-measure zero sets coincides with the class of all μ'-measure zero sets). Clearly, μ is Radon if and only if μ' is Radon. Hence we must establish that μ' is a Radon measure. Since we have

$$\mu'(D) = sup\{\mu'(C) \ : \ C \subset D \text{ and } C \text{ is closed}\}$$

for all Borel sets $D \subset F$, it suffices to demonstrate that the measure μ' is tight, i.e. for any real $\varepsilon > 0$, there exists a compact set $K(\varepsilon) \subset F$ such that

$$\mu'(F \setminus K(\varepsilon)) \leq \varepsilon.$$

Fix $\varepsilon > 0$ and take an arbitrary natural number n. Applying the separability of F, we can find a finite family

$$\{C_{n,1}, C_{n,2}, ..., C_{n,k(n)}\}$$

of closed balls in F, such that

$$diam(C_{n,i}) \leq \frac{1}{n+1} \qquad (i = 1, 2, ..., k(n));$$

$$\mu'(F \setminus (C_{n,1} \cup C_{n,2} \cup ... \cup C_{n,k(n)})) < \frac{\varepsilon}{2^{n+1}}.$$

Now, we define

$$K(\varepsilon) = \cap_{n \in \mathbf{N}} (C_{n,1} \cup C_{n,2} \cup ... \cup C_{n,k(n)}).$$

Then it is not hard to check that $K(\varepsilon)$ is a totally bounded closed subset of F (consequently, it is compact) and

$$\mu'(F \setminus K(\varepsilon)) \leq \sum_{n \in \mathbf{N}} \frac{\varepsilon}{2^{n+1}} \leq \varepsilon.$$

This completes the proof of the theorem.

Note that Theorem 1 can be extended to a much more general class of topological spaces. For instance, it remains true for the so-called analytic (or Suslin) spaces and also for co-analytic spaces. Actually, Theorem 1 may be regarded as a particular case of the fundamental Choquet theorem on capacities (see, e.g, [51], [55] or [155]).

Let F be a topological vector space and let H be a subset of F^* separating the points of F. We recall that the H-weak topology in F is defined by its local base at zero, consisting of sets of the form

$$\{x \in F \ : \ |h_1(x)| < \varepsilon, \ |h_2(x)| < \varepsilon, ..., \ |h_m(x)| < \varepsilon\}$$

where $(h_1, h_2, ..., h_m)$ is a finite subfamily of H and $\varepsilon > 0$. In these terms, the following statement is true.

Theorem 2. *Let F be a Polish topological vector space and let H be a subset of F^* separating the points of F. Let μ be an H-cylindrical premeasure on F having the property that, for each $\varepsilon > 0$, there exists an H-weakly compact set $C_\varepsilon \subset F$ such that*

$$\mu(D) < \varepsilon$$

for all sets $D \in S(F, H)$ whose intersection with C_ε is empty. Then μ admits a unique extension to a Borel measure on F.

The proof of Theorem 2 can be carried out by using the standard method. First of all, the same argument as in the proof of the classical Kolmogorov theorem (cf. [121], [155], [171], [180]) shows us that μ possesses a unique extension to a measure on the cylindrical σ-algebra $B(F, H)$. The second (and final) step of the proof is then based on one general result given in Exercise 1 of this section.

Theorem 2 can be efficiently applied in those situations where we have a visual geometric description of compact sets with respect to the weak topology. For instance, if F is a separable reflexive Banach space, then we may take various closed balls as C_ε (because closed balls in F are F^*-weakly compact sets). We may also take closed bounded convex subsets of F instead of closed balls (by the same reason). Moreover, we may deal only with compact sets C_ε in the original strong topology of F. Indeed, according to the result of Ulam presented above, every finite Borel measure on F is Radon, i.e. for any $\varepsilon > 0$, there exists a compact set $C_\varepsilon \subset F$ such that

$$\mu(F \setminus C_\varepsilon) < \varepsilon$$

or, equivalently,

$$\mu(D) < \varepsilon$$

for all sets $D \in B(F)$ whose intersection with C_ε is empty. We thus see that, for our F, the assumptions of Theorem 2 are sufficient and necessary simultaneously.

Probability cylindrical premeasures on the algebra $S(F, H)$ of a topological vector space F, where $H \subset F^*$, can successfully be defined by using various families of mutually consistent (compatible) probability measures given in finite-dimensional product-spaces. This important approach was created by Kolmogorov (see [121]), and his famous theorem mentioned earlier is basic for such an approach. In order to describe it, we take an arbitrary injective finite family $(h_1, h_2, ..., h_m)$ of functionals from H and fix a Borel probability measure $\mu_{h_1,h_2,...,h_m}$ on the Euclidean space $\mathbf{R}^m$. The family of these probability measures is said to be consistent (compatible) if:

(1) $\mu_{h_1,h_2,...,h_m}(D) = \mu_{h_1,h_2,...,h_m,h_{m+1}}(D \times \mathbf{R})$ for any injective family $(h_1, h_2, ..., h_m, h_{m+1})$ of functionals from H and for all Borel subsets D of $\mathbf{R}^m$;

(2) $\mu_{h_1,h_2,...,h_m}(D) = \mu_{h_{i_1},h_{i_2},...,h_{i_m}}(\phi(D))$ for all Borel subsets D of $\mathbf{R}^m$ and for the transformation ϕ of $\mathbf{R}^m$ induced by the permutation of coordinates associated with a permutation $(i_1, i_2, ..., i_m)$ of $(1, 2, ..., m)$.

Now, we define a mapping

$$\mu : S(F, H) \to [0, 1]$$

by the formula

$$\mu(C_{h_1,h_2,...,h_m}(D)) = \mu_{h_1,h_2,...,h_m}(D)$$

for each cylindrical set $C_{h_1,h_2,...,h_m}(D)$ from $S(F, H)$. It can directly be checked that μ is a probability premeasure on $S(F, H)$. It is also clear that this correspondence between probability cylindrical premeasures and families of mutually consistent probability measures

$$(\mu_{h_1,h_2,...,h_m})_{\{h_1,h_2,...,h_m\} \subset H}$$

is bijective and canonical. More general versions of the Kolmogorov theorem are usually formulated in terms of projective systems of measures and are discussed, e.g., in [31] and [171].

These widely known facts were recalled in order to emphasize the following circumstance.

Let μ be a Borel probability measure on the product space $\mathbf{R}^{\mathbf{N}}$ and let G be a Borel vector subspace of this space. If $\mu(G) = 1$, then we may treat μ as a probability measure on $(G, B_1(G))$ where $B_1(G)$ denotes the

σ-algebra in G induced by the Borel σ-algebra of $\mathbf{R}^{\mathbf{N}}$ (or, equivalently, induced by the cylindrical σ-algebra generated by $(pr_n)_{n\in\mathbf{N}}$). Further, if G is endowed with some topology compatible with the vector structure of G, then, in general,

$$B_1(G) \neq B(G)$$

where $B(G)$ denotes the Borel σ-algebra of G. The reader can check this fact by considering various simple examples (it suffices to take $G = l_\infty$).

Suppose for a moment that

$$B_1(G) = B(G).$$

Then, evidently, μ may be regarded as a Borel probability measure on the space G.

Let G be again a Borel vector subspace of $\mathbf{R}^{\mathbf{N}}$ and let μ be a probability cylindrical premeasure on $S(G, H)$ where $H = (pr_n)_{n\in\mathbf{N}}$. As we know, μ is uniquely determined by some family of mutually consistent Borel probability measures

$$(\mu_{j_1,j_2,\dots,j_m})_{\{j_1,j_2,\dots,j_m\}\subset\mathbf{N}}$$

which are defined on finite-dimensional Euclidean spaces. Obviously, we may consider the latter family as an H-cylindrical probability premeasure μ' on $\mathbf{R}^{\mathbf{N}}$ and hence (in view of Example 8) as a Borel probability measure on $\mathbf{R}^{\mathbf{N}}$ (which also is denoted by u'). Assume now that G is of full measure with respect to μ' (or, equivalently, that $u'(G) = 1$). In this case we also say that G is a support of μ'. Then, according to the remark above, μ' turns out to be a probability measure on the σ-algebra $B(G, H)$ and the restriction of μ' to the algebra $S(G, H)$ coincides with the original premeasure u. We thus conclude that, in the case of a Borel vector subspace G of $\mathbf{R}^{\mathbf{N}}$, the question of extending a premeasure given on the algebra $S(G, H)$ to a measure on the σ-algebra $B(G, H)$ is equivalent to the question whether G is a set of full measure (with respect to the corresponding Borel probability measure on $\mathbf{R}^{\mathbf{N}}$).

If, in addition, the equality $B(G, H) = B(G)$ holds (as, e.g., for $G = c_0$), then the approach just described is sometimes more preferable than Theorem 2.

We have discussed typical constructions of probability cylindrical measures in topological vector spaces. According to these constructions, we first define a cylindrical premeasure and then extend this premeasure on the corresponding cylindrical σ-algebra. Of course, there are many other methods for constructing probability measures. Some of them will be useful below. Let us briefly outline several constructions of different type.

Let (G, S, μ) be a measure space, let (G', S') be a measurable space and let

$$\phi : G \to G'$$

be a measurable mapping. We define a measure μ_ϕ on the σ-algebra S' by the formula

$$\mu_\phi(Z) = \mu(\phi^{-1}(Z)) \qquad (Z \in S').$$

The measure μ_ϕ is usually called the (homomorphic) image of μ with respect to the given mapping ϕ. This measure is also denoted by $\phi(\mu)$. Quite frequently, we write

$$\mu_\phi = \mu \circ \phi^{-1}.$$

Obviously, if μ is a probability measure, then μ_ϕ is a probability measure, too. This feature remains valid for all finite measures but fails to be true for σ-finite measures (cf. Exercise 8).

Example 9. For each natural number n, let μ_n be a probability measure on the space $(\mathbf{R}, B(\mathbf{R}))$. Then the family $(\mu_n)_{n \in \mathbf{N}}$ yields the probability product measure

$$\mu = \prod_{n \in \mathbf{N}} \mu_n$$

defined on the space $(\mathbf{R}^{\mathbf{N}}, B(\mathbf{R}^{\mathbf{N}}))$. Consider an arbitrary infinite matrix

$$A = (a_{nk})_{n \in \mathbf{N},\ k \in \mathbf{N}}$$

satisfying the relations:

$$a_{nk} \in \mathbf{R} \qquad (n \in \mathbf{N},\ k \in \mathbf{N}),$$

$$a_{nk} = 0 \qquad (k > n).$$

This triangular matrix induces a certain Borel linear mapping (operator) acting from $\mathbf{R}^{\mathbf{N}}$ into itself. Indeed, we may put

$$A(x) = \Big(\sum_{0 \le k \le n} a_{nk} x_k \Big)_{n \in \mathbf{N}}$$

for any element $x = (x_k)_{k \in \mathbf{N}}$ from $\mathbf{R}^{\mathbf{N}}$ (for the sake of simplicity, we denote this mapping by the same symbol A). Then the measure

$$\mu_A = \mu \circ A^{-1}$$

is the image of μ under the Borel linear mapping A and hence is a Borel probability measure on $\mathbf{R}^{\mathbf{N}}$. Changing arbitrarily $(\mu_n)_{n \in \mathbf{N}}$ and A, we obtain some class of probability measures which will be denoted below by

$$M = M(\mathbf{R}^{\mathbf{N}}).$$

Certain subclasses of M are of interest from the probabilistic viewpoint and will be considered in the sequel (see, for instance, Sections 10 and 11).

New measures can also be obtained by using the operation of convolution which plays a remarkable role in various questions of analysis and probability theory. We recall that if μ_1 and μ_2 are any two Borel measures on a Polish topological vector space F (or, more generally, on a commutative Polish topological group F), then their convolution $\mu = \mu_1 * \mu_2$ is defined by the formula

$$\mu(D) = \int_F \mu_1(D - x)\mu_2(dx) \qquad (D \in B(F)).$$

Clearly, if μ_1 and μ_2 are probability measures, then their convolution μ is a probability measure, too. Moreover, the convolution operation is commutative, so the family of all Borel probability measures in F turns out to be a commutative semigroup with respect to this operation.

Various properties of the convolution operation, important from the analytic viewpoint, are discussed in many books (cf. e.g., [124] and [175]).

Now, let us briefly consider some fundamental concepts of probability theory. According to the classical axiomatic approach (see [121], [139]), the original object of this theory is a basic probability space

$$(\Omega, S, P),$$

i.e. Ω is a nonempty set (set of elementary events), S is a σ-algebra of subsets of Ω (the σ-algebra of events) and P is a probability measure on S. For any event $A \in S$, the value $P(A)$ is usually called the probability of A.

In many cases below, it will be convenient to assume that a probability space (Ω, S, P) is complete (i.e. the corresponding probability measure P is complete, which means that all subsets of an arbitrary P-measure zero set are P-measurable). In fact, this assumption does not restrict the generality of considerations, because we may always replace any probability measure by its completion.

Let an event $A \in S$ be such that $P(A) = 1$. Then we say that this event is fulfilled almost surely, or is valid with P-probability one, or is satisfied for P-almost all $\omega \in \Omega$, etc.

Some fairly standard probabilistic terminology will be utilized in the sequel, too.

Fix a basic probability space (Ω, S, P) and let (F, S') be some measurable space. A random element in F (more precisely, an F-valued random element) is any mapping

$$X : \Omega \rightarrow F$$

measurable with respect to the σ-algebras S and S'. Below, various topological vector spaces will be taken as F and the corresponding cylindrical σ-algebras in F will be taken as S'. The case when S' coincides with the Borel σ-algebra $B(F)$ is of especial interest. In this case, a random element

$$X : (\Omega, S) \rightarrow (F, B(F))$$

is a measurable mapping in the usual sense, and we say that X is a strong random element in F (or, briefly, that X is a random element in F). Further, if $H \subset F^*$ and X is a measurable mapping from (Ω, S) into $(F, B(F, H))$, then we say that X is an H-weak random element in F (if H is fixed under our consideration, then we simply say that X is a weak random element in F).

For concrete spaces F, we use more special terminology.

For example, if $F = \mathbf{R}$, then we say that an F-valued random element is a random variable. If

$$F = \mathbf{R} \cup \{-\infty, +\infty\},$$

then we say that an F-valued random element is an extended random variable. Also, if

$$F = \mathbf{R}^m \qquad (m > 1),$$

then we say that an F-valued random element is a random vector.

Example 10. Let F be a Polish topological vector space and let H be a subset of F^* separating the points of F. Then each H-weak random element turns out to be a random element in F (in this connection, see Exercise 1).

Let us consider an arbitrary set T of indices (parameters). Any family $(X_t)_{t \in T}$ of random variables defined on (Ω, S, P) will be called a random (stochastic) process.

If $T = \mathbf{N}$, then the corresponding stochastic process is called a random sequence.

Example 11. Let a stochastic process $(X_t)_{t\in T}$ be given. Then we may associate to it a certain H-weak random element X in the product space $\mathbf{R}^T$, where $H = (pr_t)_{t\in T}$. Namely, we put

$$X(\omega) = (X_t(\omega))_{t\in T} \qquad (\omega \in \Omega).$$

Conversely, if X is some H-weak random element in $\mathbf{R}^T$, then the family $(X_t)_{t\in T}$, where

$$X_t = pr_t \circ X \qquad (t \in T),$$

turns out to be a stochastic process. Thus, we can conclude that a stochastic process with the set of indices T and an H-weak random element in the space $\mathbf{R}^T$ are, in fact, the same object. This conclusion is important from the methodological point of view.

In particular, for $T = \mathbf{N}$, we obtain that there is a canonical one-to-one correspondence between random sequences and random elements in $\mathbf{R}^{\mathbf{N}}$.

Also, in many cases, it is convenient to treat a stochastic process $(X_t)_{t\in T}$ as a function

$$X \ : \Omega \times T \to \mathbf{R}$$

of two variables $\omega \in \Omega$ and $t \in T$, defined by

$$X(\omega, t) = X_t(\omega) \qquad ((\omega, t) \in \Omega \times T).$$

Such a treatment is very useful because enables to introduce the concept of a measurable stochastic process (see, e.g, [55] or [155]).

Let X be a random element in a measurable space (F, S'). For each $\omega \in \Omega$, the value $X(\omega)$ is called a realization of this random element. Realizations of a random element are also called its trajectories. Let $C \in S'$. Sometimes, for the sake of convenience, we write briefly

$$\{X \in C\}$$

instead of

$$\{\omega \in \Omega \ : \ X(\omega) \in C\}.$$

Analogous reduced notation will be applied in many other similar situations (for equality relations, inclusions and so on).

Further, having a random element X in a measurable space (F, S'), we introduce a probability measure μ_X on (F, S') by the formula

$$\mu_X(C) = P(\{X \in C\}) \qquad (C \in S').$$

The measure μ_X is usually called the distribution of X in F. Evidently, μ_X turns out to be the homomorphic image of P with respect to the measurable mapping X. In other words, $\mu_X = \mu \circ X^{-1}$.

Now, let X and Y be any two random elements in F (in general, they may be defined on distinct probability spaces). We say that X and Y are equidistributed (or similar) if their distributions in F are identical, i.e.

$$\mu_X = \mu_Y.$$

In this case, we also say that Y is a copy of X (and, conversely, X is a copy of Y).

Two random elements in F defined on a probability space (Ω, S, P) are called to be stochastically equivalent (or P-equivalent, or, simply, equivalent) if

$$P(\{X \neq Y\}) = 0.$$

Clearly, stochastically equivalent random elements are equidistributed.

In many cases, stochastically equivalent random elements can be identified. Actually, probability theory deals with the equivalence classes of stochastically equivalent elements.

In our further considerations, we shall denote by $L_0((\Omega, S), F)$ the set of all classes of stochastically equivalent random elements in F. In particular, such an approach enables us to consider those random elements in F which are defined only P-almost everywhere in Ω. Obviously, we may arbitrarily extend those elements on the whole set Ω.

Let J be a set of indices and let, for each $j \in J$, a random element X_j be given on a basic probability space (Ω, S, P), with values in a measurable space (F_j, S_j). The family $(X_j)_{j \in J}$ of random elements is called (stochastically) independent if, for any finite sequence

$$\{j_1, j_2, ..., j_m\} \subset J$$

of pairwise distinct indices and for any sets

$$C_{j_1} \in S_{j_1}, \ C_{j_2} \in S_{j_2}, \ ..., \ C_{j_m} \in S_{j_m},$$

we have the equality

$$P(\cap_{1 \leq k \leq m} \{X_{j_k} \in C_{j_k}\}) = \prod_{1 \leq k \leq m} P(\{X_{j_k} \in C_{j_k}\}).$$

Example 12. Let X and Y be two independent random elements defined on (Ω, S, P), with values in some Polish topological vector space F. Then the sum

$$Z = X + Y$$

is a random element in F, too, and a direct verification shows that

$$\mu_Z = \mu_X * \mu_Y .$$

Clearly, if X is a random element in a measurable space (F, S'), then the triple

$$(F, S', \mu_X)$$

is a probability space. Conversely, if (F, S', ν) is a measurable space with a probability measure ν, then ν can be regarded as the distribution of some random element X in F. Namely, we put

$$(\Omega, S, \mu) = (F, S', \nu)$$

and define a random element $X : \Omega \to F$ by the formula

$$X(\omega) = \omega \qquad (\omega \in \Omega).$$

At first sight, such an approach is trivial and useless. But it is convenient in many cases, because enables us to treat any probability measure as the distribution of an appropriate random element.

Example 13. Let J be again a set of indices and let, for each $j \in J$, a probability measure μ_j be given on a measurable space (F_j, S_j). Consider the product measure space

$$(\Omega, S, P) = (\prod_{j \in J} F_j, \prod_{j \in J} S_j, \prod_{j \in J} \mu_j).$$

Further, for any $j \in J$, define a mapping

$$X_j \ : \ \Omega \to F_j$$

by the formula

$$X_j(\omega) = \omega_j \qquad (\omega = (\omega_j)_{j \in J} \in \Omega).$$

Then X_j is a random element in (F_j, S_j) and the equality

$$\mu_{X_j} = \mu_j$$

holds true for all $j \in J$. Moreover, it can easily be verified that the random elements X_j $(j \in J)$ are mutually independent.

The general facts of measure theory and probability theory observed in this section will be exploited in the sequel for more concrete situations interesting from the probabilistic viewpoint.

EXERCISES

1. Let F be a Polish topological vector space and let H be a subset of F^*. Show that the following two assertions are equivalent:

a) H separates the points of F;

b) $B(F, H) = B(F)$.

For the proof of this equivalence, utilize the classical result from descriptive set theory, stating that if a Borel injection

$$\phi : E \to E'$$

is given acting from a Polish space E into a metric space E', then, for each Borel subset C of E, the image $\phi(C)$ is a Borel subset of E' (see, e.g., [131]).

Generalize the above-mentioned equivalence of assertions a) and b) to the case when F is an arbitrary Polish space and H is a family of continuous real-valued functions on F, separating all points of F.

2. Consider the product space $\mathbf{R}^{\mathbf{N}}$ and take the countable family

$$H = (pr_n)_{n \in \mathbf{N}}$$

consisting of all canonical projections of this space on $\mathbf{R}$. Show that:

a) the σ-algebra $B(l_\infty, H)$ is countably generated (i.e. is generated by some countable family of subsets of l_∞);

b) the σ-algebra $B(l_\infty, (l_\infty)^*)$ is not countably generated;

c) there exists a closed subset of l_∞ which does not belong to the σ-algebra $B(l_\infty, (l_\infty)^*)$.

Deduce from these facts that all the inclusions

$$B(l_\infty, H) \subset B(l_\infty, (l_\infty)^*) \subset B(l_\infty)$$

are proper.

3. Give a detailed proof of Theorem 2 from this section.

4. Prove the following classical result of Luzin and Sierpiński.

Let E and E' be any two Polish topological spaces. Then these two assertions are equivalent:

a) $card(E) = card(E')$;

b) E and E' are Borel isomorphic, i.e. there exists a bijection ϕ from E onto E' such that both mappings ϕ and ϕ^{-1} are Borel.

Deduce from this result that if E is an uncountable Polish space, then there exists a Borel isomorphism

$$\psi \ : \ (E, B(E)) \to ([0,1], B([0,1])).$$

5. Let F be a Polish topological vector space, let H be a vector subspace of F^* separating the points of F, and let μ be a probability measure given on $(F, B(F))$. We recall that the characteristic functional ϕ_μ of μ is defined by the formula

$$\phi_\mu(h) = \int_F exp(ih(x))\mu(dx) \qquad (h \in H)$$

where i denotes the imaginary number for which $i^2 = -1$. Suppose now that ν is some probability measure on $(F, B(F))$. Demonstrate that the following two assertions are equivalent:

a) $\phi_\mu = \phi_\nu$;

b) $\mu = \nu$.

In other words, a probability measure μ on $(F, B(F)) = (F, B(F, H))$ is completely determined by its characteristic functional ϕ_μ.

6. Let F be again a Polish topological vector space, let H be a vector subspace of F^* separating the points of F and let μ be a Borel probability measure on F. Prove that μ is completely determined by its values on the half-spaces of the form

$$\{x \in F \ : \ h(x) \leq \theta\}$$

where $h \in H$ and $\theta \in \mathbf{R}$.

7. Let F_1 and F_2 be any two Polish topological vector spaces, let $H_1 \subset F_1^*$ and $H_2 \subset F_2^*$ be vector spaces separating the points of F_1 and F_2, respectively, and suppose that a probability measure μ on the product space

$$(F, B(F)) = (F_1, B(F_1)) \times (F_2, B(F_2))$$

is given. Denote

$$\mu_1 = \mu \circ pr_1^{-1}, \qquad \mu_2 = \mu \circ pr_2^{-1},$$

i.e. μ_1 and μ_2 are the projections of μ on F_1 and F_2 respectively. Denote also by ϕ_μ, ϕ_{μ_1}, ϕ_{μ_2} the corresponding characteristic functionals. Show that the following two statements are equivalent:

a) $\mu = \mu_1 \times \mu_2$;

b) for all $h_1 \in H_1$ and $h_2 \in H_2$, the equality

$$\phi_\mu(h_1, h_2) = \phi_{\mu_1}(h_1) \cdot \phi_{\mu_2}(h_2)$$

holds true, where $(h_1, h_2) \in F^*$ is defined by the formula

$$(h_1, h_2)(x, y) = h_1(x) + h_2(y) \qquad ((x, y) \in F).$$

8. Let μ be a σ-finite measure on some σ-algebra of sets. For any two μ-measurable sets A and B with $\mu(A) + \mu(B) < +\infty$, we put

$$d(A, B) = \mu(A \Delta B)$$

where Δ denotes the operation of symmetric difference of sets. Prove that the pair $(dom(\mu), d)$ is a complete metric space (as usual, we identify all those A and B for which $d(A, B) = 0$).

We say that a measure μ is separable if the above-mentioned metric space is separable (equivalently, is Polish). If this metric space is not separable, then we say that μ is a nonseparable measure.

Demonstrate that if $dom(\mu)$ is a countably generated σ-algebra, then μ is separable. On the other hand, give an example of a separable measure μ such that $dom(\mu)$ is not countably generated.

Further, by applying the construction of a product measure, give an example of a nonseparable probability measure on the space $[0, 1]^T$ where T is an uncountable set of indices.

Show also that a homomorphic image of a separable probability measure is separable, too. Finally, give an example of a σ-finite measure whose homomorphic image is not a σ-finite measure (for instance, such a measure is the standard Lebesgue measure λ_1 on $\mathbf{R}$).

In connection with the result of Exercise 8, let us point out that there exist nonseparable measures on $\mathbf{R}$ extending the Lebesgue measure λ_1. Moreover, as demonstrated by Kakutani, Oxtoby and Kodaira, there are nonseparable extensions of λ_1 invariant under the group of all isometric transformations of $\mathbf{R}$. For more information around this topic, see e.g. [113] and [117].

9. Show that there exist a probability space (Ω, S, P) and a topological probability space $(E, B(E), \mu)$, such that there is no random element

$$f : \Omega \to E$$

for which $\mu = \mu_f$, where μ_f denotes the distribution of f.

On the other hand, let

$$(\Omega, S, P) = ([0,1], B([0,1]), \lambda)$$

where λ is the standard Lebesgue probability measure on $[0,1]$, and let $(E, B(E), \mu)$ be a Polish topological space equipped with a Borel probability measure μ vanishing on all one-element subsets of E. Prove (by using the result of Exercise 4 above) that there exists a random element

$$g : \Omega \to E$$

which is a Borel isomorphism and whose distribution coincides with μ, i.e. $\mu = P \circ g^{-1}$.

10. Let E be a set, let S be a σ-algebra of subsets of E and let μ be a measure on S. We say that μ satisfies the countable chain condition (or μ satisfies the Suslin condition) if there is no uncountable disjoint family $\{A_j : j \in J\} \subset S$ such that

$$(\forall j \in J)(\mu(A_j) > 0).$$

Demonstrate that:

a) every σ-finite measure on S satisfies the countable chain condition;

b) any homomorphic image of a measure satisfying the countable chain condition satisfies this condition, too.

11. Let (Ω, S, P) be a probability space. For any two extended (i.e. with values in $\mathbf{R} \cup \{-\infty, +\infty\}$) random variables f_1 and f_2, we put

$$f_1 \preceq f_2 \Leftrightarrow (f_1(\omega) \leq f_2(\omega) \text{ } for \text{ } almost \text{ } all \text{ } \omega \in \Omega).$$

Obviously, $\preceq$ is a preordering in the set of all extended random variables. As usual, we identify equivalent extended random variables, so the above-mentioned preordering becomes an ordering (this procedure is a very particular case of the standard factorization with respect to a given equivalence relation). Let F denote the family of all equivalence classes of extended random variables. Show that the pair $(F, \preceq)$ is a complete lattice, i.e. for any family $\{f_i : i \in I\} \subset F$, there exist $sup\{f_i : i \in I\}$ in F and $inf\{f_i : i \in I\}$ in F.

12. Let (Ω, S) be a measurable space. For any two probability measures P_1 and P_2 defined on the σ-algebra S, we put

$$P_1 \preceq P_2 \Leftrightarrow (\forall Z \in S)(P_1(Z) \leq P_2(Z)).$$

Clearly, the relation $\preceq$ is an ordering on the family M of all probability measures on (Ω, S). By using the result of the previous exercise and applying the Radon-Nikodym theorem (see, e.g., [31], [59], [86], [124], [155]), demonstrate that $(M, \prec)$ is a Dedekind complete lattice. This means that, for each nonempty family $\{P_i \ : \ i \in I\} \subset M$ such that

$$(\exists P \in M)(\forall i \in I)(P_i \leq P),$$

there exists $sup\{P_i \ : \ i \in I\}$ in M and, analogously, for each nonempty family $\{Q_i \ : \ i \in I\} \subset M$ such that

$$(\exists Q \in M)(\forall i \in I)(Q \leq Q_i),$$

there exists $inf\{Q_i \ : \ i \in I\}$ in M.

13. Prove the formula of Example 12 for the distribution of the sum of two independent random elements in a Polish topological vector space.

14. Let F be a Polish topological vector space, let H be a subspace of F^* separating the points of F, and let X_1 and X_2 be two random elements in F (defined on the same basic probability space). Denote by μ_1 and μ_2 the distributions (in F) of X_1 and X_2 respectively. Also, denote by $\mu_{1,2}$ the distribution (in $F \times F$) of the random element (X_1, X_2). Show that the following statements are equivalent:

a) X_1 and X_2 are independent;

b) $\mu_{1,2} = \mu_1 \times \mu_2$;

c) for any two functionals $h \in H$ and $g \in H$, the random variables $h(X_1)$ and $g(X_2)$ are independent.

7. Convergence of random elements

There are various kinds of convergence of random elements, important from the probabilistic viewpoint. Here we recall some of them. The main attention will be paid to the following types of convergence of random elements:

1) convergence almost surely (or convergence with probability one);
2) convergence in probability;
3) convergence in distribution.

It should be noted that, from the functional point of view, the convergence almost surely coincides with the convergence P-almost everywhere, and the convergence in probability coincides with the convergence in measure P (here and below the triple (Ω, S, P) denotes, as usual, the basic probability space on which our random elements are given).

For the sake of simplicity, we shall deal only with those random elements whose values belong to some fixed separable Banach space F. In other words, our random elements are assumed to be F-valued. Of course, analogous notions of convergence can be introduced for more general topological vector spaces (however, we do not need in the sequel such general situations).

So, let $(F, ||\cdot||)$ be a separable Banach space and suppose that our basic probability space (Ω, S, P) is complete (this assumption does not restrict essentially the generality of our considerations). Let $(Y_n)_{n\in\mathbf{N}}$ be a sequence of F-valued random elements on (Ω, S, P).

Definition 1. We say that the sequence $(Y_n)_{n\in\mathbf{N}}$ converges almost surely (or converges with probability one) if there exists a mapping

$$Y : \Omega \rightarrow F$$

such that

$$P(\{\omega \in \Omega : lim_{n\rightarrow\infty}||Y_n(\omega) - Y(\omega)|| = 0\}) = 1$$

or, briefly,

$$P(\{lim_{n\rightarrow\infty}||Y_n - Y|| = 0\}) = 1.$$

In this case, the mapping Y necessarily is an F-valued random element and, moreover, is unique almost surely (i.e. Y is unique up to the canonical P-equivalence in the family of all F-valued random elements on (Ω, S, P)). We also say that Y is a limit (almost surely or with probability one) of the given sequence $(Y_n)_{n\in\mathbf{N}}$.

Since our space F is complete, the convergence almost surely of $(Y_n)_{n\in\mathbf{N}}$ turns out to be equivalent to the almost sure Cauchy property of $(Y_n)_{n\in\mathbf{N}}$, i.e.

$$P(\{lim_{n,m\to\infty}||Y_n - Y_m|| = 0\}) = 1.$$

Definition 2. We say that $(Y_n)_{n\in\mathbf{N}}$ converges in probability if there exists an F-valued random element Y such that, for each $\varepsilon > 0$, we have

$$lim_{n\to\infty}P(\{||Y_n - Y|| \geq \varepsilon\}) = 0.$$

In this case we also say that Y is a limit (in probability) of the given sequence $(Y_n)_{n\in\mathbf{N}}$. This limit is unique (almost surely).

Taking into account the completeness of F, we see that the convergence in probability of $(Y_n)_{n\in\mathbf{N}}$ is equivalent to the Cauchy property in probability: for any $\varepsilon > 0$, the equality

$$lim_{n,m\to\infty}P(\{||Y_n - Y_m|| \geq \varepsilon\}) = 0$$

must be fulfilled.

Recall that if a sequence $(Y_n)_{n\in\mathbf{N}}$ converges almost surely, then it converges in probability as well. The converse assertion is not true, in general. Nevertheless, it is well known that if a sequence $(Y_n)_{n\in\mathbf{N}}$ converges in probability, then it contains some subsequence which converges almost surely (see, e.g., [31], [154], [155]).

The relation

$$||h(x)|| \leq ||h|| \cdot ||x|| \qquad (h \in F^*, \ x \in F)$$

obviously implies that if a sequence $(Y_n)_{n\in\mathbf{N}}$ converges in probability, then, for each functional $h \in F^*$, the sequence of random variables

$$(h(Y_n))_{n\in\mathbf{N}} = (h \circ Y_n)_{n\in\mathbf{N}}$$

converges in probability, too. The converse statement is not true, in general, but it holds under an additional natural assumption. In order to demonstrate this fact, let us introduce one important notion from topological measure theory.

Definition 3. Let F be a topological space and let $(\mu_j)_{j\in J}$ be a family of Borel probability measures on F. We say that this family is tight if, for any real $\varepsilon > 0$, there exists a compact set $K_\varepsilon \subset F$ such that

$$sup_{j\in J}\ \mu_j(F \setminus K_\varepsilon) < \varepsilon.$$

If F is a Polish topological space and all μ_j from our family coincide with some fixed Borel probability measure μ on F, then the family is tight in virtue of the classical Ulam theorem (see the preceding section). The same remains true for any family $(\mu_j)_{j\in J}$ consisting of finitely many Borel probability measures on F. Moreover, by applying the Ulam theorem, it is not hard to infer that a sequence $(\mu_n)_{n\in\mathbf{N}}$ of Borel probability measures on F is tight if and only if, for each real $\varepsilon > 0$, there exists a compact set $K_\varepsilon \subset F$ satisfying the relation

$$limsup_{n\to\infty}\mu_n(F \setminus K_\varepsilon) < \varepsilon.$$

Now, let F be a separable Banach space and let a sequence $(Y_n)_{n\in\mathbf{N}}$ of F-valued random elements be convergent in probability to an F-valued random element Y. Denote:

$\mu_n =$ the distribution of Y_n $(n \in \mathbf{N})$;

$\mu =$ the distribution of Y.

It can be shown that the sequence of probability measures $(\mu_n)_{n\in\mathbf{N}}$ is tight. Indeed, fix an arbitrary $\varepsilon > 0$ and take any decreasing sequence of strictly positive real numbers $(\delta_l)_{l\in\mathbf{N}}$, such that

$$lim_{l\to\infty}\delta_l = 0.$$

Since $(Y_n)_{n\in\mathbf{N}}$ converges in probability, we may find, for every natural number l, a natural number $n(\delta_l, \varepsilon)$ such that

$$sup_{n>n(\delta_l,\varepsilon)}P(\{||Y_n - Y|| \geq \delta_l\}) < \varepsilon 2^{-(l+2)}.$$

For any set $A \subset F$ and for any $\delta > 0$, let us put

$$A^{(\delta)} = \{x \in F\ :\ inf_{y\in A}||x - y|| \leq \delta\}.$$

According to the Ulam theorem, for each $l \in \mathbf{N}$, there exist compact sets

$$K_j(\varepsilon, l) \qquad (j = 0, 1, ..., n(\delta_l, \varepsilon) + 1)$$

in F such that

$$P(\{Y \notin K_0(\varepsilon, l)\}) = \mu(F \setminus K_0(\varepsilon, l)) < \varepsilon 2^{-(l+2)},$$

$$sup_{j\leq n(\delta_l,\varepsilon)}\mu_j(F\setminus K_{j+1}(\varepsilon,l))<\varepsilon 2^{-(l+1)}.$$

Consequently, for each $l\in\mathbf{N}$, we obtain

$$sup_{n>n(\delta_l,\varepsilon)}\mu_n(F\setminus(K_0(\varepsilon,l))^{(\delta_l)})=sup_{n>n(\delta_l,\varepsilon)}P(\{Y_n\notin(K_0(\varepsilon,l))^{(\delta_l)}\})$$

$$\leq P(\{Y\notin K_0(\varepsilon,l)\})+sup_{n>n(\delta_l,\varepsilon)}P(\{||Y_n-Y||>\delta_l\})<\varepsilon 2^{-(l+1)}$$

and

$$sup_{n\in\mathbf{N}}\mu_n(F\setminus\cup_{j\leq n(\delta_l,\varepsilon)+1}(K_j(\varepsilon,l))^{(\delta_l)})\leq$$

$$max(sup_{n\leq n(\delta_l,\varepsilon)}\mu_n(F\setminus K_{n+1}(\varepsilon,l)),sup_{n>n(\delta_l,\varepsilon)}\mu_n(F\setminus(K_0(\varepsilon,l))^{(\delta_l)}))$$

$$<\varepsilon 2^{-(l+1)}.$$

Evidently, the set

$$K_\varepsilon=\cap_{l\geq 0}(\cup_{j\leq n(\delta_l,\varepsilon)+1}(K_j(\varepsilon,l))^{(\delta_l)})$$

is closed and totally bounded in the space F. Hence, K_ε turns out to be a compact subset of F. Besides, we may write

$$sup_{n\in\mathbf{N}}\mu_n(F\setminus K_\varepsilon)\leq\sum_{l\in\mathbf{N}}sup_{n\in\mathbf{N}}\mu_n(F\setminus\cup_{0\leq j\leq n(\delta_l,\varepsilon)+1}(K_j(\varepsilon,l))^{(\delta_l)})<\varepsilon,$$

which implies that the sequence of probability measures $(\mu_n)_{n\in\mathbf{N}}$ is tight.

Now, we are able to formulate the following statement (criterion) concerning the convergence in probability of F-valued random elements (cf. [37] and the comments therein).

Theorem 1. *Let F be a separable Banach space and let H be a subset of F^* separating the points of F. Suppose also that $(Y_n)_{n\in\mathbf{N}}$ is a sequence of F-valued random elements and let, for each natural number n, the symbol μ_n denote the distribution of Y_n. Then $(Y_n)_{n\in\mathbf{N}}$ is convergent in probability if and only if the next two relations are satisfied:*

1) the sequence of measures $(\mu_n)_{n\in\mathbf{N}}$ is tight;

2) for any $h\in H$, the sequence of random variables $(h(Y_n))_{n\in\mathbf{N}}$ converges in probability.

Note that if $(Y_n)_{n\in\mathbf{N}}$ converges in probability, then relations 1) and 2) are obviously fulfilled. Thus, it remains to demonstrate that these relations are sufficient for the convergence in probability of $(Y_n)_{n\in\mathbf{N}}$. In order to show this, we need two auxiliary propositions.

For each $h\in F^*$ and for each $t\in\,]0,1[$, we denote

$$C_t(h)=\{x\in F\ :\ |h(x)|>t||x||\}.$$

Evidently, the set $C_t(h)$ is open and does not contain the neutral element of F. Let us observe that if

$$(x_n)_{n\in\mathbf{N}} \subset C_t(h),$$

then

$$lim_{n\to\infty}||x_n|| = 0 \Leftrightarrow lim_{n\to\infty}|h(x_n)| = 0.$$

Lemma 1. *Let a set $H \subset F^*$ separate the points of a separable Banach space F and let K' be a compact subset of F such that $0 \notin K'$. Then there exist a finite set $H(K') \subset H$ and a number $t(K') \in]0,1[$, such that*

$$K' \subset \cup_{h\in H(K')} C_{t(K')}(h).$$

Proof. Since H separates the points of F, we have

$$K' \subset F \setminus \{0\} = \cup_{h\in H} \cup_{0<t<1} C_t(h).$$

Therefore, the open sets

$$C_t(h) \qquad (h \in H,\ 0 < t < 1)$$

form a covering of K'. Taking into account the compactness of K', we can find a finite subcovering

$$(C_{t_1}(h_1), C_{t_2}(h_2), ..., C_{t_m}(h_m))$$

of K'. It remains to put

$$H(K') = (h_1, h_2, ..., h_m),$$

$$t(K') = min(t_1, t_2, ..., t_m)$$

and to observe that

$$C_{t_j}(h_j) \subset C_{t(K')}(h_j)$$

for all $j \in \{1, 2, ..., m\}$. This ends the proof of Lemma 1.

Lemma 2. *Let F be again a separable Banach space and let $H \subset F^*$ separate the points of F. Then, for any compact set $K \subset F$ and for each real $\varepsilon > 0$, there exist a finite set $H(K, \varepsilon) \subset H$ and a number $t = t(K, \varepsilon) \in]0, 1[$, such that*

$$P(\{||X|| \geq \varepsilon\}) \leq P(\{X \notin K\}) + \sum_{h\in H(K,\varepsilon)} P(\{|h(X)| \geq t\varepsilon\})$$

for every F-valued random element X.

Proof. Fix $\varepsilon > 0$ and take any compact subset K of F. Clearly, we have

$$P(\{||X|| \geq \varepsilon\}) \leq P(\{||X|| \geq \varepsilon\} \cap \{X \in K\}) + P(\{X \notin K\}).$$

Further, the set

$$K' = K \cap \{x \in F \ : \ ||x|| \geq \varepsilon\}$$

is compact and does not contain zero, so we may apply Lemma 1 to K'. According to this lemma, there exist a finite set

$$H(K') = H(K, \varepsilon) \subset H$$

and a number

$$t = t(K') = t(K, \varepsilon) \in]0, 1[,$$

such that

$$K' \subset \cup_{h \in H(K')} C_{t(K')}(h).$$

Consequently,

$$P(\{||X|| \geq \varepsilon\} \cap \{X \in K\}) \leq \sum_{h \in H(K,\varepsilon)} P(\{||X|| \geq \varepsilon\} \cap \{X \in C_t(h)\}).$$

It remains to mention that

$$\{||X|| \geq \varepsilon\} \cap \{X \in C_t(h)\} \subset \{|h(X)| \geq t\varepsilon\}$$

for any functional $h \in F^*$. This completes the proof of Lemma 2.

Now, we are ready to establish Theorem 1. As pointed out, it suffices to demonstrate that conditions 1) and 2) of the theorem imply the convergence in probability of a given sequence $(Y_n)_{n \in \mathbf{N}}$. First of all, in virtue of 1), the sequence of distributions $(\mu_n)_{n \in \mathbf{N}}$ is tight. Therefore, for each real number $\delta > 0$, there exists a compact set $D_{\delta/2} \subset F$ satisfying the relation

$$sup_{n \in \mathbf{N}} P(\{Y_n \notin D_{\delta/2}\}) = sup_{n \in \mathbf{N}} \mu_n(F \setminus D_{\delta/2}) < \delta/2.$$

Let us put

$$K_\delta = D_{\delta/2} - D_{\delta/2}.$$

Then we easily get

$$sup_{n,m \in \mathbf{N}} P(\{Y_n - Y_m \notin K_\delta\}) < \delta.$$

Now, applying Lemma 2 to the random elements

$$X_{nm} = Y_n - Y_m \qquad (n \in \mathbf{N},\ m \in \mathbf{N})$$

and to the compact set $K = K_\delta$, and utilizing simultaneously condition 2), we infer that, for any real numbers $\delta > 0$ and $\varepsilon > 0$, the relation

$$limsup_{n,m\to\infty} P(\{||Y_n - Y_m|| > \varepsilon\}) < \delta$$

is fulfilled. Consequently, for each $\varepsilon > 0$, we obtain

$$limsup_{n,m\to\infty} P(\{||Y_n - Y_m|| > \varepsilon\}) = 0.$$

This obviously implies that our sequence $(Y_n)_{n\in\mathbf{N}}$ converges in probability, and Theorem 1 has thus been proved.

We see that Theorem 1 yields a certain criterion of convergence in probability of a sequence of random elements in a separable Banach space. This criterion is helpful in various situations. It should be mentioned that we do not have an analogous nice criterion for the convergence almost surely. Indeed, let $(Y_n)_{n\in\mathbf{N}}$ be again a sequence of F-valued random elements, let $(\mu_n)_{n\in\mathbf{N}}$ denote the corresponding sequence of distributions and let $H \subset F^*$ separate the points of F. One can consider the next two conditions:

1) the family of probability measures $(\mu_n)_{n\in\mathbf{N}}$ is tight;

2') for any $h \in H$, the sequence of random variables $(h(Y_n))_{n\in\mathbf{N}}$ converges almost surely.

It is clear that conditions 1) and 2') are necessary for the convergence almost surely of $(Y_n)_{n\in\mathbf{N}}$. But they are not sufficient. In other words, the convergence almost surely of $(Y_n)_{n\in\mathbf{N}}$ cannot be expressed in terms of the tightness of the corresponding distributions and in terms of the convergence almost surely of $(h(Y_n))_{n\in\mathbf{N}}$ where $h \in H$.

However, for some special types of sequences $(Y_n)_{n\in\mathbf{N}}$, the convergence almost surely admits a suitable description (see, e.g., exercises to this section). In particular, such a description is possible when the random elements Y_n $(n \in \mathbf{N})$ coincide with partial sums of a series of mutually independent random elements (cf. Example 1 below).

It is also reasonable to recall here the notion of weak convergence (for Borel probability measures) and the notion of convergence in distribution (for random elements).

Definition 4. Let F be a metric space and let $(\mu_n)_{n\in\mathbf{N}}$ be a sequence of Borel probability measures on F. We say that this sequence is weakly convergent if there exits a Borel probability measure μ on F such that, for any bounded continuous function

$$g : F \to \mathbf{R},$$

we have

$$lim_{n\to\infty} \int_F g(x)\mu_n(dx) = \int_F g(x)\mu(dx).$$

Further, in accordance with this definition, we say that a family $(\mu_j)_{j\in J}$ of Borel probability measures on F is weakly relatively compact if any infinite sequence of measures from $(\mu_j)_{j\in J}$ contains a weakly convergent subsequence.

The next classical result (due to Prokhorov [168]) establishes a remarkable relationship between the weak relative compactness and the tightness of a family of Borel probability measures.

Theorem 2. *Let $(\mu_j)_{j\in J}$ be a family of Borel probability measures on a metric space F. Then the tightness of this family implies its weak relative compactness. Conversely, if the given space F is Polish, then each weakly relatively compact family $(\mu_j)_{j\in J}$ of Borel probability measures on F turns out to be tight.*

For the proof, see e.g. [14], [80] or [168]. More general versions of this theorem can be found in [31].

Definition 5. Let $(Y_n)_{n\in\mathbf{N}}$ be a sequence of F-valued random elements and let $(\mu_n)_{n\in\mathbf{N}}$ be the sequence of corresponding distributions. We say that $(Y_n)_{n\in\mathbf{N}}$ converges in distribution if $(\mu_n)_{n\in\mathbf{N}}$ is weakly convergent.

Note that some analogue of Theorem 1 in terms of convergence in distribution is valid (see Theorem 3 below).

It is well known that the classical notion of the characteristic function of a Borel probability measure defined on a finite-dimensional Euclidean space plays the fundamental role in various questions of probability theory. This notion was generalized by Kolmogorov [123] to the case of Borel probability measures given on Banach spaces.

Namely, let μ be a Borel probability measure on a Banach space F. A complex-valued functional ϕ_μ defined on F^* by the formula

$$\phi_\mu(h) = \int_F exp(ih(x))\mu(dx) \qquad (h \in F^*),$$

where $i^2 = -1$, is said to be the characteristic functional of the measure μ. Obviously, we may write

$$\phi_\mu(h) = \int_{\mathbf{R}} exp(it)\mu_h(dt)$$

where μ_h denotes the homomorphic image of μ with respect to the measurable mapping

$$h \, : \, (F, B(F)) \to (\mathbf{R}, B(\mathbf{R})).$$

In a similar way, the characteristic functional can be introduced for cylindrical probability measures on topological vector spaces.

It is widely known that, in finite-dimensional Euclidean spaces, every Borel probability measure can be uniquely reconstructed by means of its characteristic functional. An analogous fact also holds for Borel probability measures in separable Banach spaces. Moreover (see Exercise 5 from Section 6), if μ_1 and μ_2 are any two Borel probability measures on a separable Banach space F and $H \subset F^*$ separates the points of F, then the relation

$$(\forall h \in lin(H))(\phi_{\mu_1}(h) = \phi_{\mu_2}(h))$$

implies the equality

$$\mu_1 = \mu_2.$$

By starting with this fact and taking into account the Prokhorov theorem formulated above, the following statement can easily be obtained.

Theorem 3. *Let F be a separable Banach space, let $H \subset F^*$ separate the points of F and let $(Y_n)_{n \in \mathbf{N}}$ be a sequence of random elements in F. Denote again by $(\mu_n)_{n \in \mathbf{N}}$ the sequence of corresponding distributions. Then $(Y_n)_{n \in \mathbf{N}}$ is convergent in distribution if and only if the next two conditions are satisfied:*

(1) the sequence of probability measures $(\mu_n)_{n \in \mathbf{N}}$ is tight;

(2) for each functional $h \in lin(H)$, the sequence of random variables $(h(Y_n))_{n \in \mathbf{N}}$ converges in distribution.

Comparing the formulations of Theorems 1 and 3, we see that they are very similar. This circumstance turns out to be rather helpful in various situations.

The following example illustrates the usefulness of the concepts and considerations presented above.

Example 1. Let F be a separable Banach space and let $(X_k)_{k \in \mathbf{N}}$ be a sequence of mutually independent F-valued random elements. Then the next three assertions are equivalent:

a) the series $\sum_{k\in\mathbf{N}} X_k$ converges in distribution;
b) the series $\sum_{k\in\mathbf{N}} X_k$ converges in probability;
c) the series $\sum_{k\in\mathbf{N}} X_k$ converges almost surely.

Note that, for a sequence $(Z_k)_{k\in\mathbf{N}}$ of mutually independent random variables, the equivalence of the corresponding analogues of a), b) and c) is stated by the classical Lévy theorem (see [136]). Taking this fact into account and applying Theorems 1 and 3 to the sequence of F-valued random elements $(Y_n)_{n\in\mathbf{N}}$, where

$$Y_n = \sum_{0\le k\le n} X_k \qquad (n \in \mathbf{N}),$$

we immediately get the equivalence of a) and b). The implication

$$b) \Rightarrow c)$$

is discussed in Exercise 4 to this section. Finally, it can easily be seen that the implication

$$c) \Rightarrow a)$$

holds true for any sequence $(X_n)_{n\in\mathbf{N}}$ of random elements in F. Indeed, it follows at once from the classical Lebesgue theorem on majorized convergence that, for any sequence of random elements in F, the convergence almost surely implies the convergence in distribution. The converse assertion is not true in general. In this connection, one remarkable result of Skorokhod must be pointed out:

Theorem 4. *Let F be a Polish topological space and let $(\mu_n)_{n\in\mathbf{N}}$ be a sequence of Borel probability measures on F. Suppose also that this sequence weakly converges to a Borel probability measure μ on F and consider a probability space (Ω', S', P') where*

$$\Omega' = [0,1], \quad S' = B([0,1]),$$

$$P' = \text{the standard Lebesgue measure on } [0,1].$$

Then, for this probability space, there exist a sequence of F-valued random elements $(Y_n)_{n\in\mathbf{N}}$ and an F-valued random element Y, such that:

(a) $(Y_n)_{n\in\mathbf{N}}$ converges almost surely to Y;

(b) for each natural number n, the measure μ_n coincides with the distribution of Y_n;

(c) the measure μ coincides with the distribution of Y.

For the proof of Theorem 4, see [182].

Further information about convergence of random elements can be found e.g. in [14], [28], [37], [57], [81], [103], [136], [162], [168], [197].

EXERCISES

1. Let F be an infinite-dimensional separable Hilbert space. Demonstrate that there exists a sequence $(X_n)_{n\in\mathbf{N}}$ of F-valued random elements, satisfying the following relations:

a) for each vector $f \in F$, the sequence $(< f, X_n >)_{n\in\mathbf{N}}$ of random variables converges to zero almost surely (here $<,>$ denotes the inner product in F);

b) the sequence $(X_n)_{n\in\mathbf{N}}$ converges in probability to zero;

c) the sequence $(X_n)_{n\in\mathbf{N}}$ is divergent almost surely.

2. Give a detailed proof of Theorem 3 from this section.

3. Let F be a Polish topological vector space and let $X_1, X_2, ..., X_n$ be mutually independent random elements in F. Suppose also that a Borel set $C \subset F$ is given for which

$$C - C \in B(F)$$

and

$$sup_{1\leq m\leq n} P(\{\sum_{m\leq k\leq n} X_k \notin C\}) \leq \alpha < 1.$$

Show that the following Lévy-Octaviani inequality holds true:

$$P(\bigcup_{1\leq m\leq n} \{\sum_{1\leq k\leq m} X_k \notin C - C\}) \leq \frac{\alpha}{1-\alpha}.$$

In particular, for a separable Banach space $(F, ||\cdot||)$, consider this inequality in the case when C is an open ball in F, i.e.

$$C = \{x \in F : ||x|| < r\}$$

where r is a strictly positive real number.

4. By applying the Lévy-Octaviani inequality (see the preceding exercise) to some open balls in a separable Banach space F, prove the implication

$$b) \Rightarrow c)$$

from Example 1 of this section.

5. Let F be a topological vector space and let X be a random element in F. We say that X is symmetric if, for any Borel set $D \subset F$, the equality

$$P(\{X \in D\}) = P(\{-X \in D\})$$

holds true.

Now, let $(F, ||\cdot||)$ be a separable Banach space, let H be a vector subspace of F^* separating the points of F, and let $(X_k)_{k \in \mathbf{N}}$ be a sequence of mutually independent symmetric F-valued random elements. For each natural number n, we denote by μ_n the distribution of the random element

$$Y_n = \sum_{0 \leq k \leq n} X_k.$$

Demonstrate that if there exists a compact set C (in the H-weak topology) for which we have

$$limsup_{n \to \infty} \mu_n(C) > 0,$$

then the sequence $(Y_n)_{n \in \mathbf{N}}$ converges almost surely in F.

In connection with Exercise 5, see also [37].

6. Let F be a metric space, let $(\mu_n)_{n \in \mathbf{N}}$ be a sequence of Borel probability measures on F and let μ be a Borel probability measure on the same space. Show that the following two statements are equivalent:

a) the sequence $(\mu_n)_{n \in \mathbf{N}}$ weakly converges to μ;

b) $lim_{n \to \infty} \mu_n(D) = \mu(D)$ for every Borel set $D \subset F$ satisfying the relation $\mu(bd(D)) = 0$ where $bd(D)$ denotes the boundary of D.

7. Check that Theorem 1, Theorem 3 and the statement of Example 1 remain true in a more general situation where a separable Banach space F is replaced by a Polish topological vector space F_1 such that F_1^* separates the points of F_1.

8. The structure of supports of Borel measures

This section is devoted to topological and geometrical aspects of the theory of σ-finite Borel measures given on metric spaces. Namely, here we investigate the structure of supports of such measures. Since every nonzero σ-finite measure is equivalent to a probability measure (defined on the same σ-algebra), we may restrict our further considerations to the case of probability measures.

Let E be a topological space and let μ be a Borel measure on E. We recall that a Borel set $D \subset E$ is a support of μ (or, in other words, μ is concentrated on D) if

$$\mu(E \setminus D) = 0.$$

Obviously, for a Borel probability measure μ on E, the following two assertions are equivalent:

1) D is a support of μ;

2) $\mu(D) = 1$.

For our further purposes, we need one important result concerning supports of σ-finite Borel measures given on a metric space E whose topological weight is a nonmeasurable cardinal number (see the precise definition below). We recall that the topological weight of E is the smallest cardinality among the cardinalities of bases of E. It is usually denoted by the symbol $w(E)$.

For example, taking the classical topological vector spaces

$$\mathbf{R}^{\mathbf{N}}, \quad c_0, \quad l_\infty,$$

we have respectively

$$w(\mathbf{R}^{\mathbf{N}}) = w(c_0) = card(\mathbf{N}),$$

$$w(l_\infty) = card(\mathbf{R}).$$

Actually, the last equality was demonstrated in Section 5.

An infinite cardinal number a is real-valued measurable if there exists a nonzero σ-finite measure defined on the σ-algebra of all subsets of a and vanishing on all its one-element subsets (obviously, in this definition we may consider any set A with $card(A) = a$, instead of a). Some information about real-valued measurable cardinals is presented in Appendix 3. Here we only wish to point out that real-valued measurable cardinals are so large that their existence cannot be established within the theory **ZFC** (i.e. the Zermelo-Fraenkel set theory with the Axiom of Choice). So, in order to obtain the existence of such cardinals, we are forced to appeal to additional set-theoretical axioms.

We first formulate and prove the following auxiliary statement from the theory of metric spaces (cf. [131] and [159]).

Lemma 1. *Let (E, d) be a metric space, let I be a set of indices linearly ordered by some relation $\preceq$, and let*

$$(A_i)_{i \in I}, \quad (B_i)_{i \in I}, \quad (C_i)_{i \in I}$$

be three families of subsets of E, such that:

1) all sets A_i $(i \in I)$ are open in E;

2) for each index $i \in I$, we have

$$B_i = A_i \setminus \cup\{A_j \ : \ j \prec i\};$$

3) for each index $i \in I$, the set C_i is closed in E and $C_i \subset B_i$.

Then the set $\cup\{C_i \ : \ i \in I\}$ can be represented as the union of some countable family of closed subsets of E (in other words, $\cup\{C_i \ : \ i \in I\}$ is an F_σ-set in E).

Proof. For any index $i \in I$ and for any natural number $n \geq 1$, we put

$$C_i^n = \{z \ : \ z \in C_i, \ d(z, E \setminus A_i) \geq 1/n\}.$$

Evidently, we have the equality

$$\cup\{C_i^n \ : \ n \in \mathbf{N}, \ n \geq 1\} = C_i.$$

Furthermore, each set C_i^n is closed in E. Now, if

$$j \prec i, \quad x \in C_i^n, \quad y \in C_j^n,$$

then $d(x, y) \geq 1/n$. Consequently, we have

$$d(C_i^n, C_j^n) \geq 1/n \qquad (n \in \mathbf{N} \setminus \{0\})$$

for any two distinct indices i and j from I (here we utilize the assumption that the set I is linearly ordered by $\preceq$). Thus, for each natural number $n \geq 1$, the set

$$\cup\{C_i^n \ : \ i \in I\}$$

turns out to be closed in E. It remains to observe that

$$\cup\{C_i \ : \ i \in I\} = \cup\{\cup\{C_i^n \ : \ i \in I\} \ : \ n \in \mathbf{N} \setminus \{0\}\},$$

which shows that $\cup\{C_i \ : \ i \in I\}$ is an F_σ-set in E.

Now, we are able to prove the following statement (cf. [14], [146] and [159]).

Theorem 1. *Let (E, d) be a metric space whose topological weight is not real-valued measurable. Let μ be an arbitrary σ-finite Borel measure on E. Then there exists a separable support of μ, i.e. there exists a closed separable subset $F = F_\mu$ of E such that*

$$\mu(E \setminus F) = 0.$$

Proof. Since every nonzero σ-finite measure is equivalent to a probability measure, we may suppose without loss of generality that our μ is a Borel probability measure on E. Fix a natural number k and consider a covering

$$(U_{k,i})_{i \in I(k)}$$

of E by open balls whose diameters are less than or equal to $1/(k+1)$. Evidently, we can assume that

$$card(I(k)) = w(E).$$

Moreover, applying the Zermelo theorem, we can also assume that the set $I(k)$ is well ordered by some ordering relation $\preceq$. Further, for each index $i \in I(k)$, we denote

$$Z_{k,i} = U_{k,i} \setminus \cup\{U_{k,j} \ : \ j \prec i\}.$$

Clearly, the family of sets

$$\{Z_{k,i} \ : \ i \in I(k)\}$$

is disjoint and, simultaneously, is a covering of E. In addition, each set $Z_{k,i}$ (being a difference of two open sets) admits the representation

$$Z_{k,i} = \cup\{Z_{k,i,m} \ : \ m \in \mathbf{N}\}$$

where all sets $Z_{k,i,m}$ are closed in E. Since μ is a probability measure, the set of all those indices $i \in I(k)$ for which $\mu(Z_{k,i}) > 0$ is at most countable. Let us denote this set by $T(k)$ and put

$$J(k) = I(k) \setminus T(k).$$

We are going to demonstrate that

$$\mu(\cup\{Z_{k,i} \ : \ i \in J(k)\}) = 0.$$

Suppose otherwise, i.e.

$$\mu(\cup\{Z_{k,i} \ : \ i \in J(k)\}) > 0.$$

Then, for each subset J of $J(k)$, we may write

$$\cup\{Z_{k,i} \ : \ i \in J\} = \cup\{Z_{k,i,m} \ : \ i \in J, \ m \in \mathbf{N}\}.$$

According to Lemma 1, the set

$$Z_{k,m} = \cup\{Z_{k,i,m} \ : \ i \in J\}$$

is Borel in E, hence the set

$$\cup\{Z_{k,i} \ : \ i \in J\} = \cup\{Z_{k,m} \ : \ m \in \mathbf{N}\}$$

turns out to be Borel, too. Now, putting

$$\nu(J) = \mu(\cup\{Z_{k,i} \ : \ i \in J\})$$

for each set $J \subset J(k)$, we come to the nonzero finite measure ν defined on the σ-algebra of all subsets of $J(k)$ and vanishing on all one-element subsets of $J(k)$, i.e. $card(J(k))$ is a real-valued measurable cardinal. But this is impossible since

$$card(J(k)) \leq card(I(k)) = w(E)$$

and since $w(E)$ is not real-valued measurable (cf. Appendix 3). The contradiction obtained shows that

$$\mu(\cup\{Z_{k,i} \ : \ i \in J(k)\}) = 0.$$

This implies, in particular, that our measure μ is concentrated on the set

$$U(k) = \cup\{U_{k,i} \ : \ i \in T(k)\}$$

for each natural number k. Now, it can easily be checked that the set

$$D = \cap\{U(k) \ : \ k \in \mathbf{N}\}$$

is a separable subspace of E and μ is concentrated on D as well. Thus, it suffices to put $F = cl(D)$ where $cl(D)$ denotes the closure of D. The proof of Theorem 1 is complete.

Some applications of this theorem will be given later (cf. Exercises 2 and 3).

Remark 1. The assumption that a metric space E in Theorem 1 has non-real-valued measurable topological weight is very essential for the validity of this theorem. Indeed, suppose that a real-valued measurable cardinal exists and take any such a cardinal a. Further, choose any set E with $card(E) = a$ and equip E with the discrete topology (or, equivalently, with the discrete metric). Then E becomes a metrizable space whose topological weight is equal to a. Let μ be a nonzero σ-finite measure defined on the family of all subsets of E and vanishing on all one-element subsets of E. Then, obviously, μ is a Borel measure on E which does not possess a separable support (since each separable subset of E is at most countable). In other words, the assertion of Theorem 1 fails to be true for our μ.

In addition, let us notice that, for the validity of Theorem 1, the assumption of metrizability of E is essential, too.

We thus see that, under some natural assumption, any Borel probability measure on a metric space is concentrated on a separable subspace. Furthermore, if we deal with a Borel probability measure defined on a separable metric space, then the question arises whether there exists a support of this measure with nice geometric properties. Since every separable metric space is isometrically contained in a separable Banach space (more precisely, according to the classical Banach-Mazur theorem [10], every separable metric space can be realized as an isometric copy of some subset of the function space $C[0,1]$), it is reasonable to ask whether each Borel probability measure on a separable Banach space admits a support with good geometric properties. We are going to show that there always exists a support which simultaneously is a separable reflexive Banach space with a smooth norm. In order to do it, we first recall some auxiliary notions and facts from the theory of Banach spaces.

Let $(F_1, ||\cdot||_1)$ and $(F_2, ||\cdot||_2)$ be two Banach spaces. We shall say that $(F_1, ||\cdot||_1)$ is compactly embedded into $(F_2, ||\cdot||_2)$ if the following conditions are fulfilled:

1) $F_1 \subset F_2$;

2) the identity mapping from $(F_1, ||\cdot||_1)$ into $(F_2, ||\cdot||_2)$ is continuous;

3) each bounded subset of $(F_1, ||\cdot||_1)$ is relatively compact in $(F_2, ||\cdot||_2)$.

Evidently, 3) implies 2). So, in this definition, 2) can be omitted.

Remark 2. Taking into account the fact mentioned in Exercise 1 from Section 6, we see that if $(F_1, ||\cdot||_1)$ is separable and compactly embedded into $(F_2, ||\cdot||_2)$, then the inclusion

$$B((F_1, ||\cdot||_1)) \subset B((F_2, ||\cdot||_2))$$

holds true, i.e. any Borel set in F_1 remains Borel one in F_2.

Let $(F, ||\cdot||)$ be a fixed Banach space.

We say that a Banach space $(F_+, ||\cdot||_+)$ belongs to the class $U(F)$ if the next two relations are satisfied:

a) $(F_+, ||\cdot||_+)$ is compactly embedded into $(F, ||\cdot||)$;

b) $(F_+, ||\cdot||_+)$ is separable and reflexive.

We say that a Banach space $(F_+, ||\cdot||_+)$ belongs to the class $V(F)$ if it belongs to $U(F)$ and, in addition, the following relation holds:

c) the norm $||\cdot||_+$ is Frechet differentiable and locally uniformly convex.

In connection with the last relation, we recall that the differentiability (in the Frechet sense) of the norm $||\cdot||_+$ is equivalent to the condition: for any element

$$x_0 \in S_1(0) = \{x \in F_+ \ : \ ||x||_+ = 1\},$$

there exists a limit

$$lim_{t\to 0}\frac{||x_0 + tx||_+ - ||x_0||_+}{t}$$

uniformly with respect to $x \in S_1(0)$. Accordingly, the norm $||\cdot||_+$ is locally uniformly convex if

$$(\forall x \in S_1(0))(\forall (x_n)_{n\in\mathbf{N}} \subset S_1(0))(lim_{n\to\infty}\frac{||x_n + x||_+}{2} = 1 \Rightarrow$$

$$lim_{n\to\infty}||x_n - x||_+ = 0).$$

Let us emphasize that a space F_+ from the class $U(F)$ (respectively, from $V(F)$) is not, in general, a closed subset of F. Moreover, various situations are possible when the closure (in F) of F_+ coincides with F.

The definition of the class $V(F)$ shows that spaces belonging to this class possess rather good geometric properties. It will be demonstrated below that if a tight family of Borel probability measures on F is given, then there always exists a space from $V(F)$ which is a common support for all measures of the family. More precisely, we can formulate the following statement (see [40]; an analogous result holds true even for Frechet separable spaces).

Theorem 2. *Let F be a separable Banach space and let $(\mu_i)_{i\in I}$ be a tight family of Borel probability measures on F. Then there exists a space $F_+ \in V(F)$ such that*

$$(\forall i \in I)(\mu_i(F_+) = 1),$$

and the family $(\mu_i)_{i\in I}$ is also tight in F_+.

Before giving a proof of Theorem 2, let us mention two immediate consequences of this theorem (other nontrivial consequences will be pointed out later).

Theorem 2 and the Ulam classical theorem presented in Section 6 yield at once the following proposition.

(1). Let μ be a Borel probability measure on a separable Banach space F. Then there exists a space $F_+(\mu) \in V(F)$ such that μ is a Borel probability measure on $F_+(\mu)$.

Further, Theorem 1 and the result just formulated imply the next proposition.

(2). Let μ be a Borel probability measure on a Banach space F whose topological weight is not real-valued measurable. Then there exists a space $F_+(\mu) \in V(F)$ such that μ is a Borel probability measure on $F_+(\mu)$.

Now, let us start our argument leading to the proof of Theorem 2. It is based on several auxiliary statements.

Lemma 2. *Let $(F, ||\cdot||)$ be a Banach space and let $(K_n)_{n\in\mathbf{N}}$ be an arbitrary sequence of compact subsets of F. Then there exist a symmetric (with respect to zero) convex compact set $D \subset F$ and a sequence $(p_n)_{n\in\mathbf{N}}$ of strictly positive real numbers, such that*

$$K_n \subset p_n D \qquad (n \in \mathbf{N}).$$

Proof. We may assume, without loss of generality, that all sets K_n are compact, symmetric and convex (otherwise, it suffices to replace each K_n by the closed convex hull of the set $(-K_n) \cup K_n$). Also, we may suppose that

$$K_n \setminus \{0\} \neq \emptyset \qquad (n \in \mathbf{N}).$$

Further, for any $n \in \mathbf{N}$, let us put

$$d_n = max\{||x|| \ : \ x \in K_n\}, \quad p_n = (n+1)d_n, \quad C_n = (p_n)^{-1}K_n.$$

Clearly, for all $n \in \mathbf{N}$, we have $d_n > 0$, the set C_n is a symmetric convex compact subset of F and, besides,

$$(*) \quad max\{||x|| \ : \ x \in C_n\} \leq \frac{1}{n+1}.$$

Let us denote

$$C = \cup_{n \in \mathbf{N}} C_n$$

and observe that $0 \in C_n$ for each $n \in \mathbf{N}$. In addition,

$$K_n = p_n C_n \subset p_n C \qquad (n \in \mathbf{N}).$$

It is not hard to check that C is a compact subset of F. Indeed, let $(U_j)_{j \in J}$ be an arbitrary covering of C by open sets. Since $0 \in C$, there exists at least one index $j_0 \in J$ such that $0 \in U_{j_0}$. Hence, in virtue of (*), there exists a natural number n_0 for which we have

$$\cup_{n > n_0} C_n \subset U_{j_0}.$$

Further, in view of the compactness of the set $\cup_{n \leq n_0} C_n$, there are some sets $U_{j_1}, U_{j_2}, ..., U_{j_m}$ from our covering, such that

$$\cup_{n \leq n_0} C_n \subset \cup\{U_{j_k} \ : \ 1 \leq k \leq m\}.$$

Therefore,

$$C \subset \cup\{U_{j_k} \ : \ 0 \leq k \leq m\},$$

which establishes the compactness of C. Now, it can easily be seen that the set

$$D = cl(conv(C))$$

is the required one, and the proof is complete.

Lemma 2 implies at once the next statement.

Lemma 3. *Let $(F, || \cdot ||)$ be a Banach space and let $(\mu_i)_{i \in I}$ be a tight family of Borel probability measures on F. Then there exists a symmetric convex compact set $D \subset F$ for which we have*

$$lim_{r \to +\infty}(inf_{i \in I} \ \mu_i(rD)) = 1.$$

Lemma 4. *Let $(F, || \cdot ||)$ be a separable Banach space and let $(\mu_i)_{i \in I}$ be a tight family of Borel probability measures on F. Then there exists a* *separable Banach space* $(F_1, ||\cdot||_1)$ *compactly embedded into* $(F, ||\cdot||)$ *and satisfying the relation*

$$(\forall i \in I)(\mu_i(F_1) = 1).$$

Moreover, the family $(\mu_i)_{i\in I}$ *is also tight in* $(F_1, ||\cdot||_1)$.

Proof. Here we only sketch the argument, leaving some purely technical details to the reader. First, recall once more the classical Banach-Mazur theorem stating that the space $C[0,1]$ equipped with the standard norm

$$||x||_\infty = sup_{t\in[0,1]}|x(t)| \qquad (x \in C[0,1])$$

is universal for the class of all separable metric spaces. Hence our F is contained in $C[0,1]$ as its closed subset. In virtue of this result, it suffices to establish Lemma 4 only in the case when

$$(F, ||\cdot||) = (C[0,1], ||\cdot||_\infty).$$

According to Lemma 3, there exists a symmetric convex compact subset D of $C[0,1]$ for which we have

$$lim_{r\to+\infty}(inf_{i\in I}\ \mu_i(rD)) = 1.$$

Furthermore, applying the classical Ascoli-Arzelá theorem on the structure of compact subsets of $C[0,1]$ (see, e.g., [124]), we may even suppose that

$$D = \{x \in C[0,1]\ :\ |x(0)| + sup_{t\neq s}\frac{|x(t)-x(s)|}{\rho(t,s)} \leq 1\}$$

where ρ denotes an appropriate metric on $[0,1]$.

Consider now the subspace Λ_ρ of $C[0,1]$ defined by the formula

$$\Lambda_\rho = \{x \in C[0,1]\ :\ sup_{t\neq s}\frac{|x(t)-x(s)|}{\rho(t,s)} < +\infty\}.$$

Note that Λ_ρ turns out to be a Banach space with respect to the norm

$$||x||_\rho = |x(0)| + sup_{t\neq s}\frac{|x(t)-x(s)|}{\rho(t,s)} \qquad (x \in \Lambda_\rho).$$

Obviously, D coincides with the closed unit ball of the space $(\Lambda_\rho, ||\cdot||_\rho)$ and this space is compactly embedded into $(C[0,1], ||\cdot||_\infty)$. However, we cannot assert that $(\Lambda_\rho, ||\cdot||_\rho)$ is separable and that the family $(\mu_i)_{i\in I}$ is tight in it. So we are forced to appeal to another space $\Lambda^0_{\rho_1}$ defined by the equality

$$\Lambda^0_{\rho_1} = \{x \in C[0,1] \ : \ lim_{t \to s} \frac{|x(t) - x(s)|}{\rho_1(t,s)} = 0 \ for \ all \ s \in [0,1]\},$$

where ρ_1 is any metric on $[0,1]$ inducing the standard topology of $[0,1]$ and satisfying the condition

$$lim_{t \to s} \frac{\rho(t,s)}{\rho_1(t,s)} = 0 \qquad (s \in [0,1]).$$

For example, we may put

$$\rho_1(t,s) = \sqrt{\rho(t,s)} \qquad (t \in [0,1], \ s \in [0,1]).$$

Now, the space $(\Lambda^0_{\rho_1}, ||\cdot||_{\rho_1})$ is complete and separable with respect to the norm

$$||x||_{\rho_1} = |x(0)| + sup_{t \neq s} \frac{|x(t) - x(s)|}{\rho_1(t,s)} \qquad (x \in \Lambda^0_{\rho_1}).$$

Also, it can easily be verified that the embeddings

$$(\Lambda_\rho, ||\cdot||_\rho) \subset (\Lambda^0_{\rho_1}, ||\cdot||_{\rho_1}) \subset (C[0,1], ||\cdot||_\infty)$$

are compact. Consequently, the set

$$D_1 \ = \ the \ closure \ of \ D \ in \ the \ space \ (\Lambda^0_{\rho_1}, ||\cdot||_{\rho_1})$$

is compact in $(\Lambda^0_{\rho_1}, ||\cdot||_{\rho_1})$ and, in view of the choice of D, we have

$$lim_{r \to +\infty}(inf_{i \in I} \ \mu_i(rD_1)) = 1.$$

The last formula immediately implies that the family of probability measures $(\mu_i)_{i \in I}$ is tight in the space $(\Lambda^0_{\rho_1}, ||\cdot||_{\rho_1})$, so we may put

$$(F_1, ||\cdot||_1) = (\Lambda^0_{\rho_1}, ||\cdot||_{\rho_1}).$$

This ends the proof of Lemma 4.

For our further purposes, we also need one auxiliary proposition which, in fact, is the main part of the well-known factorization theorem due to Davis, Figiel, Johnson and Pelczyński (see [54] and [56]). In order to formulate it, let us recall some simple notions from the theory of Banach spaces.

Let $(Z, ||\cdot||_Z)$ be a Banach space (not necessarily separable), let B_Z be the closed unit ball in Z and let W be a symmetric convex bounded subset of Z. Let us put

$$U_n = 2^n W + 2^{-n} B_Z \qquad (n \in \mathbf{N})$$

and, for each natural number n, consider the Minkowski functional $||\cdot||_n$ associated with the set U_n. Obviously, $||\cdot||_n$ is a norm on Z equivalent to the original norm $||\cdot||_Z$. We also define

$$|||x||| = (\sum_{n\in\mathbf{N}} ||x||_n^2)^{1/2} \qquad (x \in Z),$$

$$Y = \{x \in Z \ : \ |||x||| < +\infty\}.$$

Evidently, Y is a vector subspace of Z equipped with the norm $|||\cdot|||$. Denote by B_Y the closed unit ball in $(Y, |||\cdot|||)$, and let j be the canonical embedding of Y into Z. In these terms, the following proposition is true.

Lemma 5. *The next three relations are satisfied:*

1) $W \subset B_Y$;

2) $(Y, |||\cdot|||)$ *is a Banach space and* j *is a continuous embedding of* $(Y, |||\cdot|||)$ *into* $(Z, ||\cdot||_Z)$;

3) if W *is relatively weakly compact, then the Banach space* $(Y, |||\cdot|||)$ *turns out to be reflexive.*

For the proof, see [54] or [56].

The following statement (also necessary for Theorem 2) concerns the so-called renorming operation in the class of all Banach spaces. The main goal of statements of this kind is to establish the conditions under which a given Banach space admits a new equivalent norm with certain smoothness properties. Namely, we need the well-known John-Zizler result (see [56], [100], [101]) which will be formulated here in some simplified form.

Lemma 6. *Any reflexive Banach space admits a renorming in such a way that its new norm becomes locally uniformly convex and differentiable in the Frechet sense.*

Now, we are ready to present a proof of Theorem 2.

Let $(F, ||\cdot||)$ be a separable Banach space and let $(\mu_i)_{i\in I}$ be a tight family of Borel probability measures on F. First of all, applying repeatedly Lemma 4, we can find two separable Banach spaces $(F_1, ||\cdot||_1)$ and $(F_2, ||\cdot||_2)$, such that the embeddings

$$(F_2, ||\cdot||_2) \subset (F_1, ||\cdot||_1) \subset (F, ||\cdot||)$$

will be compact and the given family of probability measures $(\mu_i)_{i \in I}$ will be tight in each of these spaces. Let us put:

$(Z, ||\cdot||_Z) = (F_1, ||\cdot||_1)$;

W = the closure (in $(F_1, ||\cdot||_1)$) of the set B_{F_2}, where B_{F_2} denotes the closed unit ball in the space $(F_2, ||\cdot||_2)$.

Since the set W is compact in the space $(F_1, ||\cdot||_1)$, we may apply Lemma 5 and, according to this lemma, there exists a reflexive Banach space $(Y, |||\cdot|||)$ such that the embeddings

$$(F_2, ||\cdot||_2) \subset (Y, |||\cdot|||) \subset (F_1, ||\cdot||_1)$$

are continuous. Let F_+ be the closure of F_2 with respect to the norm $|||\cdot|||$. Recall that a Banach space is reflexive if and only if its closed unit ball is compact in the weak topology (see, e.g., [59]). This fact implies that any closed vector subspace of a reflexive Banach space is reflexive, too. So we infer that $(F_+, |||\cdot|||)$ is a reflexive Banach space. Moreover, we can assert that $(F_+, |||\cdot|||)$ is separable because of the separability of $(F_2, ||\cdot||_2)$. Further, from the definition of $(F_+, |||\cdot|||)$ directly follows that it is compactly embedded into $(F, ||\cdot||)$, i.e. belongs to the class $U((F, ||\cdot||))$. In addition, the family of probability measures $(\mu_i)_{i\in I}$ turns out to be tight in $(F_+, |||\cdot|||)$ (since this family is tight in $(F_2, ||\cdot||_2)$ and $(F_2, ||\cdot||_2)$ is continuously embedded into $(F_+, |||\cdot|||)$). It remains to utilize Lemma 6, according to which the norm $|||\cdot|||$ can be replaced by some equivalent norm $||\cdot||_+$ in such a way that the space $(F_+, ||\cdot||_+)$ becomes an element of the class $V((F, ||\cdot||))$. Theorem 2 has thus been proved.

Earlier, two immediate consequences of Theorem 2 were formulated. Let us briefly discuss other consequences of this theorem which are interesting from the probabilistic point of view. All of them are closely connected with various types of convergence of probability measures and random elements (the corresponding notions were introduced in Section 7).

(3). Let $(F, ||\cdot||)$ be a separable Banach space and let $(\mu_n)_{n\in\mathbf{N}}$ be a weakly convergent sequence of Borel probability measures on this space. Then there exists a space $(F_+, ||\cdot||_+) \in V((F, ||\cdot||))$ such that $(\mu_n)_{n\in\mathbf{N}}$ weakly converges in $(F_+, ||\cdot||_+)$, too.

The proof follows at once from Theorem 2 above and Theorem 3 of Section 7 by applying the trivial fact that F^* separates the points of F_+.

Now, consider an arbitrary family of F-valued random elements $(Y_i)_{i\in I}$ where $(F, ||\cdot||)$ is again a separable Banach space (here and below, we assume that all Y_i are defined on the same basic probability space). Theorem

2 shows us that if the family of probability distributions $(\mu_{Y_i})_{i\in I}$ is tight in $(F, ||\cdot||)$, then $(Y_i)_{i\in I}$ may also be regarded as a family of random elements in some space $(F_+, ||\cdot||_+)$ belonging to the class $V((F, ||\cdot||))$. Moreover, we see that the family of distributions of these random elements is also tight in the new space. The most simple example of such a situation can be obtained in the case when all Y_i $(i \in I)$ are equidistributed (actually, in this case we have only one tight probability distribution).

Further, consider a sequence of F-valued random elements $(Y_n)_{n\in \mathbf{N}}$, where $(F, ||\cdot||)$ is a separable Banach space, and suppose that this sequence converges in a certain sense: in distribution, in probability, almost surely. In each of these cases, the sequence of the corresponding distributions turns out to be tight in $(F, ||\cdot||)$ (see Section 7). Utilizing Theorem 2, we may treat $(Y_n)_{n\in \mathbf{N}}$ as a sequence of random elements in a separable Banach space $(F_+, ||\cdot||_+)$ belonging to the class $V((F, ||\cdot||))$. Since the norm $||\cdot||_+$ is stronger than $||\cdot||$, a natural question arises whether it is possible to preserve the corresponding convergence of $(Y_n)_{n\in \mathbf{N}}$ for the new space $(F_+, ||\cdot||_+)$. In this connection, it should be pointed out that the convergence in distribution and the convergence in probability are preserved.

(4). Let $(Y_n)_{n\in \mathbf{N}}$ be a sequence of random elements in a separable Banach space $(F, ||\cdot||)$, convergent in distribution. Then there exists a space $(F_+, ||\cdot||_+) \in V((F, ||\cdot||))$ such that the sequence $(Y_n)_{n\in \mathbf{N}}$ of random elements in $(F_+, ||\cdot||_+)$ remains convergent in distribution.

(5). Let $(Y_n)_{n\in \mathbf{N}}$ be a sequence of random elements in a separable Banach space $(F, ||\cdot||)$ and suppose that this sequence is convergent in probability. Then there exists a space $(F_+, ||\cdot||_+) \in V((F, ||\cdot||))$ such that the sequence $(Y_n)_{n\in \mathbf{N}}$ of random elements in $(F_+, ||\cdot||_+)$ remains convergent in probability.

The proof of (4) is very similar to the proof of (3). The proof of (5) follows from Theorem 2 of this section and Theorem 1 of Section 7 by taking into account the fact that F^* separates the points of F_+.

The case of convergence almost surely is much more delicate and is discussed in exercises below (namely, see Exercises 8 and 9). However, in those cases where the convergence almost surely is closely connected with the convergence in probability, the answer to the question posed above can be easily obtained.

(6). Let $(X_k)_{k\in \mathbf{N}}$ be a sequence of mutually independent random elements in a separable Banach space $(F, ||\cdot||)$. If the series $\sum_{k\in \mathbf{N}} X_k$ converges almost surely in $(F, ||\cdot||)$, then there exists a space $(F_+, ||\cdot||_+)$ from the class $V((F, ||\cdot||))$, such that the same series converges almost surely in $(F_+, ||\cdot||_+)$.

The proof follows immediately from (5) and the result formulated in Example 1 of Section 7, according to which, for any series of independent random elements in a separable Banach space, the convergence almost surely coincides with the convergence in probability (note that X_k $(k \in \mathbf{N})$ remain independent in the new space $(F_+, ||\cdot||_+)$).

Concluding this section, let us point out once more that Theorem 2 and its consequences presented above can be extended to the case of a separable Frechet space F (see Exercises 11 and 12; cf. also [19], [20] and [157]).

Finally, let us mention that some nontrivial applications of Theorem 1 will be discussed in our further considerations.

EXERCISES

1. Let E be a topological space satisfying the following conditions:

a) there exists a countable set $D \subset E$ everywhere dense in E;

b) for each point $d \in D$, the one-element set $\{d\}$ is a G_δ-subset of E (i.e. $\{d\}$ can be represented as the intersection of a countable family of open sets in E).

Let μ be any σ-finite Borel measure on E vanishing on D (i.e. for each $d \in D$, we have $\mu(\{d\}) = 0$). Show that there exists a set $P \subset E$ such that:

1) P is an F_σ-subset of E;

2) P is of first category in E;

3) P is a support of μ.

In particular, deduce from this result that if E is an arbitrary separable metric space, then any σ-finite Borel measure in E vanishing on all one-element subsets of E is concentrated on some first category set in E.

2. Let E be a complete metric space whose topological weight is not real-valued measurable, and let μ be a Borel probability measure on E. Show that, for each Borel set $D \subset E$ and for any real number $\varepsilon > 0$, there exists a compact set $K \subset D$ satisfying the relation

$$\mu(D \setminus K) < \varepsilon.$$

Deduce from this fact that if ν is an arbitrary σ-finite Borel measure on E and D is any Borel set in E, then

$$\nu(D) = sup\{\nu(K) \ : \ K \subset D, \ K \ is \ compact\}.$$

Thus, there exists a set $D' \subset D$ such that:

a) $\nu(D \setminus D') = 0$;

b) D' is σ-compact, i.e. D' can be represented as the union of a countable family of compact subsets of E.

These results show that every σ-finite Borel measure on E is Radon and, hence, E is a Radon space.

3. Let E be a metric space having the property that its topological weight is not real-valued measurable, and let μ be a σ-finite Borel measure in E vanishing on all one-element subsets of E. Demonstrate that there exists a set $D \subset E$ satisfying the following relations:

a) D is a separable F_σ-subset of E;

b) D is of first category in E;

c) μ is concentrated on D, i.e. D turns out to be a support of μ.

4. Let $1 \leq p_1 \leq p_2 < +\infty$. Show that the embeddings

$$(l_{p_1}, ||\cdot||_1) \subset (l_{p_2}, ||\cdot||_2) \subset (c_0, ||\cdot||_\infty)$$

are continuous but are not compact.

5. Give a concrete example of an infinite-dimensional separable reflexive Banach space which is compactly embedded into the space $(c_0, ||\cdot||_\infty)$.

6. Give a detailed proof of Lemma 4. Namely, demonstrate that:

a) the normed vector space $(\Lambda_\rho, ||\cdot||_\rho)$ is a Banach space;

b) the normed vector space $(\Lambda^0_{\rho_1}, ||\cdot||_{\rho_1})$ is a separable Banach space;

c) $(\Lambda_\rho, ||\cdot||_\rho)$ is compactly embedded into $(\Lambda^0_{\rho_1}, ||\cdot||_{\rho_1})$.

7. Give a concrete example of a separable reflexive Banach space which is compactly embedded into the space $(C[0,1], ||\cdot||_\infty)$ and, in addition, is dense everywhere in the latter space (with respect to the norm $||\cdot||_\infty$).

8. Let $(F, ||\cdot||)$ be a separable Banach space and let $(Y_n)_{n\in\mathbf{N}}$ be a sequence of F-valued random elements, converging almost surely to some F-valued random element Y. Show that there exists an increasing sequence $(t_n)_{n\in\mathbf{N}}$ of positive real numbers, for which we have:

a) $lim_{n\to\infty} t_n = +\infty$;

b) $lim_{n\to\infty} t_n ||Y_n - Y|| = 0$ almost surely.

9. Let $(F, ||\cdot||)$ be again a separable Banach space and let $(Y_n)_{n\in\mathbf{N}}$ be a sequence of F-valued random elements, converging almost surely to some F-valued random element Y. Show that there exists a sequence of separable Banach spaces

$$((F_+^{(n)}, ||\cdot||_+^{(n)}))_{n\in\mathbf{N}} \subset V((F, ||\cdot||)),$$

for which all the embeddings

$$(F_+^{(n+1)}, ||\cdot||_+^{(n+1)}) \subset (F_+^n, ||\cdot||_+^n) \qquad (n \in \mathbf{N})$$

are compact and

$$lim_{n\to\infty}||Y_n - Y||_+^{(n)} = 0$$

almost surely.

10. Prove that every separable Frechet space is isomorphic to some subspace of the product of countably many copies of the standard Banach space $C[0,1]$ (cf. Exercise 8 from Section 5).

11. Let F be a separable Frechet space. We denote by the symbol $V(F)$ the class of all those separable reflexive Banach spaces which are compactly embedded into F and whose norm is locally uniformly convex and differentiable in the Frechet sense.

Show that if $(\mu_i)_{i\in I}$ is a tight family of Borel probability measures on F, then there exists a space $(F_+, ||\cdot||_+) \in V(F)$ such that

$$\mu_i(F_+) = 1 \qquad (i \in I)$$

and the family $(\mu_i)_{i\in I}$ is also tight in $(F_+, ||\cdot||_+)$.

12. By applying the result of the previous exercise, formulate and prove the corresponding generalized versions of consequences (1) - (6) of Theorem 2 in the case when a separable Frechet space is taken instead of a separable Banach space $(F, ||\cdot||)$.

9. Quasi-invariant probability measures

Let E be an arbitrary set. Any bijection

$$g : E \to E$$

is usually called a transformation of E. Evidently, all transformations of E constitute a group with respect to the composition operation. This group is sometimes denoted by the symbol $Sym(E)$. Any subgroup of $Sym(E)$ is called a group of transformations of E.

Let G be a group of transformations of E and let S be a σ-algebra of subsets of E. We say that S is invariant with respect to G (or, briefly, G-invariant) if

$$(\forall X \in S)(\forall g \in G)(g(X) \in S).$$

Suppose, in addition, that μ is a measure defined on S. We say that μ is quasi-invariant with respect to G (or, briefly, G-quasi-invariant) if

$$(\forall X \in S)(\forall g \in G)(\mu(X) = 0 \Leftrightarrow \mu(g(X)) = 0).$$

Furthermore, we say that μ is invariant with respect to G (or, briefly, G-invariant) if

$$(\forall X \in S)(\forall g \in G)(\mu(g(X)) = \mu(X)).$$

Invariant and quasi-invariant measures can be met in various fields of mathematics and are objects of extensive investigations (see, e.g., [9], [60], [61], [85], [86], [87], [90], [92], [93], [105], [113], [114], [117], [125], [127], [128], [129], [140], [144], [147], [156], [158], [160], [169], [170], [183], [197], [198], [199]). In particular, they play an essential role in probability theory. Therefore, it is reasonable to consider here some properties of such measures from the purely geometric viewpoint. Since we primarily deal with vector spaces, the main attention will be paid to Borel probability measures given in topological vector spaces and quasi-invariant with respect to groups of translations of these spaces.

In this section we recall several important examples of invariant and quasi-invariant measures. Also, we present some general constructions of such measures and demonstrate that those constructions lead us to many other nontrivial examples of invariant and quasi-invariant measures.

Undoubtedly, the most famous examples of invariant measures are the standard Borel measure and the standard Lebesgue measure on a finite-dimensional Euclidean space.

Let n be a natural number, let $\mathbf{R}^n$ denote, as usual, the n-dimensional Euclidean space and let $B(\mathbf{R}^n)$ be the σ-algebra of all Borel subsets of $\mathbf{R}^n$. The standard Borel measure on $\mathbf{R}^n$ can be defined as a (unique) measure λ_n satisfying the following three conditions:

1) $dom(\lambda_n) = B(\mathbf{R}^n)$;

2) $\lambda_n([0,1]^n) = 1$;

3) λ_n is invariant under the group of all translations of $\mathbf{R}^n$.

Since the space $\mathbf{R}^n$ is locally compact, the measure λ_n may be regarded as a very particular case of a Haar measure on a locally compact topological group (the corresponding definition of a Haar measure will be given below).

The completion of λ_n is usually called the standard Lebesgue measure on the space $\mathbf{R}^n$ (it is denoted by the same symbol λ_n). Obviously, the standard Lebesgue measure on $\mathbf{R}^n$ can be described as a smallest (with respect to inclusion) measure λ_n on $\mathbf{R}^n$ satisfying the following three relations:

(1) λ_n is a complete measure;

(2) $[0,1]^n \in dom(\lambda_n)$ and $\lambda_n([0,1]^n) = 1$;

(3) λ_n is invariant under the group of all translations of $\mathbf{R}^n$.

It should be noted that the above-mentioned relations imply the invariance of λ_n with respect to the group of all isometric transformations of $\mathbf{R}^n$ (cf. Exercise 1).

Let E be a locally compact topological group and let G be the group of all left (right) translations of E. We recall that the left (right) Haar measure on E is a measure μ satisfying these four conditions:

a) μ is not identically zero;

b) $dom(\mu) = B(E)$, i.e. μ is a Borel measure on E;

c) μ is invariant with respect to G;

d) for each compact subset K of E, we have $\mu(K) < +\infty$.

Notice that condition d) is equivalent to the following one:

d') for each point $x \in E$, there exists a neighbourhood $V(x)$ of x such that $\mu(V(x)) < +\infty$.

In fact, the latter condition says that μ is a locally finite measure.

A famous theorem due to Haar states that there exists at least one left (right) Haar measure μ on E and, in addition, μ is unique in a certain sense.

The theory of the Haar measure is thoroughly considered in various books and monographs (see, for instance, [86]). In particular, the above-mentioned theorem concerning the existence and uniqueness of the Haar measure can be found in [86]. In our book, we do not need many facts from this theory.

Let us emphasize that if E is a compact topological group, then the left and the right Haar measures are, in fact, proportional and can be identified (this result is not true, in general, even for σ-compact locally compact topological groups; the corresponding examples of essentially different left and right Haar measures on such groups are due to von Neumann and are presented, for instance, in [86]).

Let us also remark that if E is a σ-compact locally compact topological group, then a Haar measure μ on E is σ-finite and Radon. Conversely, it is not difficult to show that if a Haar measure μ on a locally compact topological group E is σ-finite, then E is σ-compact and, consequently, μ is Radon.

The case of a Polish locally compact topological group E is of especial importance. Then, since E possesses a countable base consisting of open sets with compact closures, E is simultaneously a σ-compact space. Taking account of the preceding remarks, we see that, in such a situation, the Haar measure μ on E is σ-finite and Radon (in connection with this fact, let us recall that a more general result due to Ulam holds true: every σ-finite Borel measure on a Polish topological space is Radon).

Example 1. Let G be the group of all isometric transformations of the n-dimensional Euclidean space $\mathbf{R}^n$. Let us equip G with its natural topology (which is induced by the topology of the Euclidean space of dimension n^2). Then G becomes a locally compact Polish topological group (and, moreover, it turns out to be a Lie group). Consequently, we may consider a σ-finite Haar measure μ on G. More generally, let H be an arbitrary closed subgroup of G. Then H is also a locally compact Polish topological group and we may consider a σ-finite Haar measure ν on H. Actually, ν is isomorphic (in the purely measure-theoretical sense) to the standard Borel measure given either on the real line $\mathbf{R}$, or on the one-dimensional unit torus $\mathbf{S}^1$, or on a finite subgroup of $\mathbf{S}^1$.

Let $(\Gamma, \cdot)$ be a topological group. We recall that Γ is monothetic if there exists an element $g \in \Gamma$ such that the subgroup of Γ generated by the one-element set $\{g\}$ is dense in Γ.

For instance, it can easily be checked that the torus $\mathbf{S}^1$ (considered as a commutative compact topological group) is monothetic.

Example 2. Let G be an arbitrary compact monothetic topological group equipped with the Haar probability measure μ. Let h be an element of G generating a dense subgroup of G. We denote this dense countable subgroup by H. Suppose also that ν is a Borel probability measure on G left invariant with respect to H. Then it is not hard to show that ν coincides with the Haar measure μ. Indeed, let us put

$$\phi_h(g) = hg \qquad (g \in G)$$

and consider an arbitrary continuous real-valued function p defined on G. Since ν is left invariant with respect to H, we have the equalities

$$\int_G p(\phi_h^n(g))\nu(dg) = \int_G p(h^n g)\nu(dg) = \int_G p(g)\nu(dg)$$

for every integer n. Further, let us fix an element $f \in G$. Clearly, there exists a sequence $\{n_k \ : \ k \in \mathbf{N}\}$ of integers, such that

$$lim_{k\to\infty} h^{n_k} = f.$$

Therefore, we obtain the relations

$$\int_G p(fg)\nu(dg) = \int_G lim_{k\to\infty} p(h^{n_k} g)\nu(dg) = \int_G p(g)\nu(dg)$$

which show us that, for each $f \in G$, the ν-integrals of the functions

$$p \ : \ G \to \mathbf{R},$$

$$p \circ \phi_f \ : \ G \to \mathbf{R}$$

are equal. Since p and f were taken arbitrarily, it is not difficult to deduce from the result obtained that our measure ν is left invariant with respect to the whole group G. Now, applying the uniqueness theorem to the Haar measure μ, we get the required equality $\nu = \mu$.

Let G be a Polish topological group. We know that if G is locally compact, then there exists a nonzero σ-finite Haar measure on G invariant with respect to the group of all left (right) translations of G. Suppose now that G is not locally compact. Then the following question arises naturally: does there exist a nonzero σ-finite Borel measure on G invariant with respect to the group of all left (right) translations of G? Or, if the answer is negative, one can pose the following question: does there exist a nonzero σ-finite Borel measure on G quasi-invariant with respect to the group of all left (right) translations of G?

As far as we know, the questions formulated above were first raised by Gelfand. Several authors investigated these questions, and it was established that the answers to them are negative (in this connection, see e.g. [183] and the references given there). Moreover, a much stronger result can be obtained by using some simple geometrical properties of σ-compact subsets of a non-locally compact Polish group G.

Lemma 1. *Let $(G, \cdot)$ be a topological group and let H and K be subsets of G. Suppose also that the following two conditions are fulfilled:*

1) H is not σ-compact in G (i.e. H cannot be covered by a countable family of compact subsets of G);

2) K is σ-compact in G.

Then there exists an uncountable disjoint family consisting of left (right) H-translates of K.

Proof. Obviously, it is sufficient to consider only the case when the elements of H act in G as left translations of G. Let ω_1 denote, as usual, the least uncountable ordinal number. We are going to construct, by using the method of transfinite recursion, a disjoint ω_1-family $\{K_\xi : \xi < \omega_1\}$ consisting of some left H-translates of the given set K. First of all, let us put $K_0 = K$. Suppose now that, for a nonzero ordinal number $\xi < \omega_1$, the partial disjoint family $\{K_\zeta : \zeta < \xi\}$ of left H-translates of K has already been constructed. Let us define

$$Z = \cup\{K_\zeta : \zeta < \xi\}.$$

Then, taking account of condition 2) and the inequality $card(\xi) < \omega_1$, we obtain that the set Z is σ-compact in G. Consequently, the set

$$Z \cdot Z^{-1} = \{x \cdot y^{-1} : x \in Z, \ y \in Z\}$$

is σ-compact in G, too. According to condition 1), we may write

$$H \setminus (Z \cdot Z^{-1}) \neq \emptyset.$$

Let h be an arbitrary element from $H \setminus (Z \cdot Z^{-1})$. Evidently, we have

$$hZ \cap Z = \emptyset.$$

Let us put $K_\xi = hK$. Then it is clear that the partial family $\{K_\zeta : \zeta \leq \xi\}$ is also disjoint and consists of some left H-translates of K. In such a way, we are able to define the required disjoint ω_1-family $\{K_\xi : \xi < \omega_1\}$ of left H-translates of K. Thus, the proof of Lemma 1 is complete.

This lemma immediately implies the following statement.

Theorem 1. *Let G be a Polish topological group and let H be a non-σ-compact subgroup of G. Then there exists no nonzero σ-finite left (right) H-quasi-invariant Borel measure on G.*

In particular, if G is not a locally compact group, then there exists no nonzero σ-finite left (right) G-quasi-invariant Borel measure on G.

Proof. Suppose that μ is a nonzero σ-finite left (or, respectively, right) H-quasi-invariant Borel measure on G. Since G is a Polish topological space, μ must be a nonzero Radon measure. Consequently, there exists a compact subset K of G such that $\mu(K) > 0$. Now, according to Lemma 1, there exists an uncountable disjoint family of left (right) H-translates of K. Since μ is left (right) H-quasi-invariant, all sets from the above-mentioned family have strictly positive μ-measures, which yields a contradiction with the fact that μ satisfies the countable chain condition. The contradiction obtained finishes the proof.

In connection with Theorem 1, it is reasonable to remark that some Polish non-locally compact topological groups admit nonzero left (right) invariant measures (see, e.g., [9] and [158]). Of course, those measures are not σ-finite and, therefore, they are of less interest for mathematical analysis, probability theory and their applications.

Also, we wish to point out here the fundamental result, due to Mackey [144], concerning the structure of those Borel subgroups of a Polish topological group which admit nonzero σ-finite Borel quasi-invariant measures.

Namely, let G be a Borel subgroup of some Polish topological group (quite frequently, G is called a standard group). Then the following two assertions are equivalent:

1) there exists a nonzero σ-finite Borel measure on G which is quasi-invariant with respect to the group of all left translations of G;

2) there exist a locally compact Polish topological group Γ and a group isomorphism

$$\Phi : G \rightarrow \Gamma,$$

such that Φ is simultaneously a Borel isomorphism between G and Γ.

Moreover, if μ stands for some nonzero σ-finite Borel measure on G quasi-invariant under the group of all left translations of G, then the group isomorphism Φ can be chosen in such a way that the image

$$\Phi(\mu) = \mu \circ \Phi^{-1}$$

turns out to be equivalent to the left Haar measure ν on Γ (i.e. the class of all $\Phi(\mu)$-measure zero sets coincides with the class of all ν-measure zero sets).

In particular, by starting with this deep and important result, it is easy to demonstrate, for any Polish topological group H, that the following two relations are equivalent:

(1) H is locally compact;

(2) there exists a nonzero σ-finite Borel measure on H quasi-invariant with respect to the group of all left translations of H.

A sketch of the proof of this equivalence is outlined in Exercise 2.

As mentioned earlier, the standard Borel measure on the real line $\mathbf{R}$ is an essentially unique measure defined on the σ-algebra of all Borel subsets of $\mathbf{R}$ and invariant with respect to the group of all translations of $\mathbf{R}$. By using this uniqueness property, it is not difficult to describe the class of all nonzero σ-finite $\mathbf{R}$-quasi-invariant Borel measures on $\mathbf{R}$ (see Exercise 5).

Now, let us consider some natural operations which enable us to produce new spaces with invariant (quasi-invariant) measures, starting from an initial family of spaces equipped with invariant (quasi-invariant) measures.

First, let us recall the purely algebraic notion of a product (respectively, weak product) of a family of groups.

Suppose that a family of groups $\{(G_i, \cdot_i) \ : \ i \in I\}$ is given and let e_i denote the neutral element of G_i for each index $i \in I$.

As usual, the product group

$$(G, \cdot) = (\prod_{i \in I} G_i, \cdot)$$

of this family is defined by the formula

$$g \cdot h = (g_i \cdot_i h_i)_{i \in I}$$

for any two elements $g = (g_i)_{i \in I}$ and $h = (h_i)_{i \in I}$ from $G = \prod_{i \in I} G_i$. Further, we put

$$G^* = \{g \in G \ : \ \{i \in I \ : \ g_i \neq e_i\} \ is \ finite\}.$$

It can easily be checked that G^* is a subgroup of G. The group G^* is usually called the weak product group of a given family $\{(G_i, \cdot_i) \ : \ i \in I\}$. If all groups $(G_i, \cdot_i)$ $(i \in I)$ are commutative, then the weak product group coincides with the direct sum of a given family of groups.

The next simple statement concerns the behaviour of invariant (respectively, quasi-invariant) measures under the operation of their product.

Theorem 2. *Let $\{(E_i, G_i) : i \in I\}$ be a family of spaces equipped with transformation groups. Suppose that, for each index $i \in I$, a probability measure μ_i is given on the set E_i. We denote by E the product of the family of sets $\{E_i : i \in I\}$ and by μ - the product measure of the family $\{\mu_i : i \in I\}$. Also, we denote by G the product group of the family $\{G_i : i \in I\}$ and by G^* - the weak product group of this family.*

The following two assertions hold:

1) if all measures μ_i $(i \in I)$ are G_i-invariant, then the product measure μ is G-invariant;

2) if all measures μ_i $(i \in I)$ are G_i-quasi-invariant, then the product measure μ is G^*-quasi-invariant.

Proof. Assertion 1) follows directly from the definition of the product measure and from the observation that the invariance of measure is preserved when one starts with an algebra of sets and goes to the σ-algebra generated by that algebra (let us mention that an analogous fact is not true for the quasi-invariance of measure).

Assertion 2) can easily be proved by using the classical Fubini theorem.

As an immediate application of Theorem 2, we have the next example.

Example 3. Let $\{\mu_i : i \in I\}$ be a family of probability measures, such that, for any $i \in I$, the measure μ_i is defined on the Borel σ-algebra of $\mathbf{R}$ and is $\mathbf{R}$-quasi-invariant. As usual, we denote by $\mathbf{R}^{(I)}$ the vector subspace of $\mathbf{R}^I$ coinciding with the direct sum of copies of $\mathbf{R}$ whose number is equal to $card(I)$. Obviously, $\mathbf{R}^{(I)}$ is everywhere dense in the topological vector space $\mathbf{R}^I$. According to Theorem 2, the product measure μ of the given family $\{\mu_i : i \in I\}$ is defined on the canonical cylindrical σ-subalgebra of $B(\mathbf{R}^I)$ and is quasi-invariant with respect to the vector space $\mathbf{R}^{(I)}$.

In particular, if $card(I) \leq card(\mathbf{N})$, then $dom(\mu)$ coincides with $B(\mathbf{R}^I)$ and, consequently, μ turns out to be a Borel probability measure on the space $\mathbf{R}^I$, quasi-invariant with respect to the dense vector subspace $\mathbf{R}^{(I)}$ of $\mathbf{R}^I$.

Thus, we have a large class of Borel probability measures on the infinite-dimensional Polish topological vector space $\mathbf{R}^{\mathbf{N}}$, which are quasi-invariant with respect to the dense vector subspace $\mathbf{R}^{(\mathbf{N})}$ of $\mathbf{R}^{\mathbf{N}}$. Note that $\mathbf{R}^{(\mathbf{N})}$ is, in some sense, a maximal vector subspace of $\mathbf{R}^{\mathbf{N}}$ possessing the property described in Example 3 (in this connection, see Exercise 7).

Having a Borel probability measure on $\mathbf{R}^{\mathbf{N}}$ quasi-invariant under a dense vector subspace of $\mathbf{R}^{\mathbf{N}}$, one can try to construct analogous measures for other infinite-dimensional topological vector spaces. In particular, the following question arises naturally: does there exist, for every separable Banach space E, a Borel probability measure μ on E such that μ is quasi-invariant with respect to some dense vector subspace of E? It turns out that the answer to this question is positive. In order to show this, we need one simple lemma from the theory of Banach spaces. First, let us recall that l_1 denotes, as usual, the classical Banach space consisting of all those sequences $(x_n)_{n\in\mathbf{N}} \in \mathbf{R}^{\mathbf{N}}$ for which the inequality

$$\sum_{n\in\mathbf{N}} |x_n| < +\infty$$

is fulfilled. The space l_1 contains $\mathbf{R}^{(\mathbf{N})}$, is separable and, moreover, is the initial space in the class of all infinite-dimensional separable Banach spaces. More precisely, we have the following auxiliary statement.

Lemma 2. *Let E be an arbitrary separable Banach space. Then there exists a continuous linear surjection*

$$\Phi \ : \ l_1 \rightarrow E.$$

Proof. Let us denote by $||\cdot||$ the norm in E and put

$$B = \{e \in E \ : \ ||e|| \leq 1\}.$$

Since E is separable, we can choose a sequence of points

$$\{e_n \ : \ n \in \mathbf{N}\} \subset B,$$

everywhere dense in B. Now, take an arbitrary element

$$x = (x_n)_{n\in\mathbf{N}} \in l_1$$

and define

$$\Phi(x) = \sum_{n\in\mathbf{N}} x_n e_n.$$

Note that this definition is correct because

$$\sum_{n\in\mathbf{N}} ||x_n e_n|| \leq \sum_{n\in\mathbf{N}} |x_n| < +\infty.$$

Also, it can easily be seen that Φ is a linear continuous mapping from l_1 into E. Finally, it remains to check that Φ is a surjection.

Let e be any point of B. We shall construct, by recursion, an injective infinite subfamily $\{e_{n(k)} \ : \ k \in \mathbf{N}\}$ of the sequence $\{e_n \ : \ n \in \mathbf{N}\}$, such that

$$||e - e_{n(0)}/2^0 - \ ... \ - e_{n(k)}/2^k|| \leq 1/2^{k+1}$$

for each $k \in \mathbf{N}$. Evidently, we can find an element $e_{n(0)}$ satisfying the inequality

$$||e - e_{n(0)}|| \leq 1/2.$$

Suppose now that, for a natural number k, the partial family

$$\{e_{n(0)}, \ e_{n(1)}, \ ... \ , \ e_{n(k)}\}$$

has already been defined. Then we may write

$$||2^{k+1}(e - e_{n(0)} - ... - e_{n(k)}/2^k)|| \leq 1.$$

Since the sequence $\{e_n \ : \ n \in \mathbf{N}\}$ is dense in B, we can choose an index $n(k+1)$ such that

$$n(k+1) > max\{n(0), n(1), ..., n(k)\}$$

and

$$||2^{k+1}(e - e_{n(0)} - ... - e_{n(k)}/2^k) - e_{n(k+1)}|| \leq 1/2.$$

From the latter inequality we immediately obtain

$$||e - e_{n(0)} - ... - e_{n(k)}/2^k - e_{n(k+1)}/2^{k+1}|| \leq 1/2^{k+2}.$$

In this way, we are able to construct the desired family $\{e_{n(k)} \ : \ k \in \mathbf{N}\}$. Now, let us put

$$y = (y_n)_{n \in \mathbf{N}} \in \mathbf{R}^{\mathbf{N}}$$

where $y_n = 0$ if n differs from all $n(k)$ $(k \in \mathbf{N})$, and $y_n = 1/2^k$ if $n = n(k)$ for some $k \in \mathbf{N}$. Then it is easy to see that

$$y \in l_1, \qquad \Phi(y) = e.$$

Consequently, we get the relation

$$B \subset \Phi(l_1).$$

Taking account of the linearity of Φ, we conclude that $\Phi(l_1) = E$. This finishes the proof of the lemma.

Theorem 3. *Let E be an arbitrary separable Banach space. Then there exists a Borel probability measure on E quasi-invariant with respect to some dense vector subspace of E.*

Proof. We start with a Borel probability measure ν on l_1 which is quasi-invariant with respect to a dense vector subspace of l_1 (the existence of ν follows, e.g., from Exercise 8 where a much stronger result is formulated). Further, according to Lemma 2, there exists a linear continuous surjection

$$\Phi : l_1 \to E.$$

Let us put

$$\mu(X) = \nu(\Phi^{-1}(X)) \qquad (X \in B(E)).$$

Then, as we know, μ is a Borel probability measure on E (being the image of ν under Φ). Moreover, it can easily be verified that if the original measure ν is quasi-invariant with respect to a dense vector subspace U of l_1, then the measure μ is quasi-invariant with respect to the vector subspace $V = \Phi(U)$ of E. Taking account of the fact that Φ is a continuous surjection, we see that V is everywhere dense in E. This completes the proof of Theorem 3.

In connection with the preceding result, a general problem of the existence of a Borel probability measure on a topological vector space, quasi-invariant with respect to a dense vector subspace, can be posed. More exactly, we can formulate the following problem: give a characterization of all those topological vector spaces E for which there exists at least one Borel probability measure quasi-invariant with respect to some dense vector subspace of E. This problem still remains open (for the class of all topological vector spaces). However, for the class of all Banach spaces, we may assume that the above-mentioned problem is solved. Indeed, Theorem 3 states that every separable Banach space E admits a Borel probability measure quasi-invariant with respect to a dense vector subspace of E. On the other hand, if E is a nonseparable normed vector space whose topological weight is not a real-valued measurable cardinal number, then E does not admit a Borel probability measure quasi-invariant with respect to a dense vector subspace of E (cf. Exercise 3 to this section, where a more general result is presented).

Thus, under the set-theoretical assumption that no cardinal number is real-valued measurable, we can infer that, for a given Banach space E, these two assertions are equivalent:

1) E is a separable space;

2) there exists a Borel probability measure on E quasi-invariant with respect to some dense vector subspace of E.

Fortunately, it turns out that an additional set-theoretical assumption can be omitted here (see Exercise 9 which contains the corresponding result in the theory **ZFC**).

Evidently, the problem posed above has a natural analogue for nonzero σ-finite invariant Borel measures. Namely, we can formulate the following problem: give a characterization of all those topological vector spaces E for which there exists at least one nonzero σ-finite Borel measure invariant with respect to some dense vector subspace of E. This problem remains open, too. Here, we shall establish only the fact that certain infinite-dimensional topological vector spaces admit a nonzero σ-finite Borel measure invariant under a dense vector subspace. In order to show this, we need the following general statement (see [117]).

Theorem 4. *Let E be a set, let $\{G_n : n \in \mathbf{N}\}$ be a sequence of groups of transformations of E and let $\{\mu_n : n \in \mathbf{N}\}$ be a sequence of σ-finite measures given on E. Suppose also that these five conditions are satisfied:*

1) the family $\{G_n : n \in \mathbf{N}\}$ is increasing with respect to inclusion;

2) the family $\{dom(\mu_n) : n \in \mathbf{N}\}$ is increasing with respect to inclusion;

3) for each $n \in \mathbf{N}$, the measure μ_n is invariant under the group G_n;

4) there exists a sequence $\{Y_n : n \in \mathbf{N}\}$ of subsets of E, such that, for any $n \in \mathbf{N}$, we have

$$Y_n \in dom(\mu_n), \qquad Y_n \subset Y_{n+1}$$

and Y_n is a G_n-invariant support of μ_n, i.e.

$$\mu_n(E \setminus Y_n) = 0, \qquad (\forall g \in G_n)(g(Y_n) = Y_n);$$

5) for each $n \in \mathbf{N}$, if a set X belongs to $dom(\mu_{n+1})$, then the set $X \cap Y_n$ belongs to $dom(\mu_n)$ and we have the equality

$$\mu_n(X \cap Y_n) = \mu_{n+1}(X \cap Y_n).$$

Let us denote

$$G = \cup\{G_n : n \in \mathbf{N}\}$$

and put

$$S = \{Z \subset E : (\forall n \in \mathbf{N})(Z \cap Y_n \in dom(\mu_n))\}.$$

Then S is a G-invariant σ-algebra of subsets of E. Further, for each set Z from S, we define

$$\mu(Z) = lim_{n \to \infty} \mu_n(Z \cap Y_n).$$

Then the definition of μ is correct, μ is a σ-finite G-invariant measure on the σ-algebra S and the set

$$Y = \cup\{Y_n \ : \ n \in \mathbf{N}\}$$

is a G-invariant support of μ. Furthermore, for any $n \in \mathbf{N}$, we have

$$\mu|Y_n = \mu_n|Y_n.$$

Proof. First, let us observe that if $Z \in S$, then, for all natural numbers n, the relation

$$\mu_n(Z \cap Y_n) = \mu_{n+1}(Z \cap Y_n) \leq \mu_{n+1}(Z \cap Y_{n+1})$$

is fulfilled. Consequently, there exists a limit

$$lim_{n\to\infty} \mu_n(Z \cap Y_n)$$

which will be denoted by $\mu(Z)$. In this way, we obtain a function

$$\mu \ : \ S \to \mathbf{R}.$$

Obviously, μ is a measure on S such that, for every natural number n, the restriction of μ to the set Y_n coincides with the measure $\mu_n|Y_n$. Other properties of μ are easily verified and we leave the corresponding details to the reader.

The measure μ described in Theorem 4 may be regarded as an inductive limit of the given countable family $\{\mu_n \ : \ n \in \mathbf{N}\}$ of invariant measures.

Example 4. We can directly apply Theorem 4 in concrete situations. For instance, let us consider the product space

$$E = \mathbf{R}^{\mathbf{N}} = \prod\{R_k \ : \ k \in \mathbf{N}\}$$

where $R_k = \mathbf{R}$ for all natural indices k. Furthermore, for each $k \in \mathbf{N}$, let us denote by λ_k the standard Borel measure on R_k and let ν_k be the standard Borel probability measure on the unit segment $[0, 1] \subset R_k$. Finally, we put

$$\mu_n = \prod\{\lambda_k \ : \ k \leq n\} \ \times \ \prod\{\nu_k \ : \ k > n\}$$

and

$$G_n = R_0 \times \ ... \ \times R_n \times \{0\} \times \{0\} \times \{0\} \times \ ...$$

for every natural number n. Now, it can easily be checked that the family $\{\mu_n : n \in \mathbf{N}\}$ of σ-finite measures and the family $\{G_n : n \in \mathbf{N}\}$ of transformation groups satisfy the assumptions of Theorem 4. Consequently, for the set E equipped with the group

$$G = \cup\{G_n : n \in \mathbf{N}\}$$

of its transformations, there exists a nonzero σ-finite G-invariant measure μ having the properties described in Theorem 4. Also, it is not difficult to see that $dom(\mu)$ contains the Borel σ-algebra of the space E, and

$$\mu([0,1]^{\mathbf{N}}) = 1.$$

We thus conclude that there exists a nonzero σ-finite Borel measure on the infinite-dimensional Polish topological vector space $\mathbf{R}^{\mathbf{N}}$, invariant with respect to the dense vector subspace $\mathbf{R}^{(\mathbf{N})}$.

Let us return for a while to Lemma 2 and recall that, in accordance with this lemma, for every separable Banach space E, there exists a linear continuous surjection

$$\Phi : l_1 \rightarrow E.$$

Also, Exercise 12 shows us that there exists a nonzero σ-finite Borel measure ν on l_1 invariant with respect to a dense vector subspace of l_1. Unfortunately, we cannot apply here (to Φ and ν) an argument similar to the proof of Theorem 3 in order to establish the existence of a nonzero σ-finite Borel measure on E invariant with respect to a dense vector subspace of E. The real reason of this fact can easily be discovered. Namely, we know that a homomorphic image of a σ-finite measure is not, in general, a σ-finite measure (cf. Exercise 8 from Section 6). So, in this situation, an argument analogous to the proof of Theorem 3 does not work.

For some related properties of invariant and quasi-invariant measures given on infinite-dimensional spaces equipped with transformation groups, see e.g. [9], [114], [117], [144], [158], [183], [197].

Now, we wish to return to a finite-dimensional Euclidean space and discuss one important geometric property of the classical Lebesgue measure given on this space. We mean here the so-called Steinhaus property (first established and investigated by Steinhaus in his old paper [186]).

Let λ_n denote the standard Lebesgue measure on the n-dimensional Euclidean space $\mathbf{R}^n$ and let G be the group of all isometric transformations of $\mathbf{R}^n$. Obviously, we may regard G as a locally compact Polish topological group (with respect to the natural group operation and topology). Consider an arbitrary Lebesgue measurable subset Z of $\mathbf{R}^n$ with $\lambda_n(Z) > 0$. Then it is not hard to show that there exists a neighbourhood U of the neutral element of G, such that

$$g(Z) \cap Z \neq \emptyset$$

for all $g \in U$ (this property of the set Z is usually called the Steinhaus property of Z). A more general result is presented in Exercise 13.

In particular, if we restrict our consideration to the group of all translations of $\mathbf{R}^n$ (canonically isomorphic to $\mathbf{R}^n$), then we easily obtain that there exists a neighbourhood W of zero in $\mathbf{R}^n$, such that

$$(h + Z) \cap Z \neq \emptyset$$

for any $h \in W$. In other words, we have the relation

$$W \subset Z - Z = \{x - y \ : \ x \in Z, \ \ y \in Z\}.$$

This property of Z can successfully be applied to the problem concerning the existence of Lebesgue nonmeasurable subsets of the space $\mathbf{R}^n$ (see, for instance, [86], [113], [154], [159]).

Note also that the Steinhaus property has a direct analogue for sets measurable with respect to a σ-finite Haar measure given on a σ-compact locally compact topological group (see, e.g., [86]).

Finally, it is reasonable to point out here that the Steinhaus property admits an analogue in terms of the Baire category for arbitrary topological groups. Namely, let $(G, \cdot)$ be a topological group and let Z be a subset of G having the Baire property (for the definition, see Appendix 3) and not belonging to the family of all first category subsets of G. Then the set $Z \cdot Z^{-1}$ contains a neighbourhood of the neutral element of G. This fact is sometimes called the Banach-Kuratowski-Pettis theorem (cf., for instance, [107] and [131]).

Let T be a nonempty compact metric space and let

$$u \ : \ T \rightarrow T$$

be some homeomorphism (in particular, u is a transformation of T). A natural question arises whether there exists a probability Borel measure on T invariant with respect to the group of transformations, generated by u. Evidently, the last group is cyclic and coincides with the family

$$\{u^m \ : \ m \in \mathbf{Z}\}$$

where $\mathbf{Z}$ denotes the set of all integers. We wish to end this section with the proof of the famous result, due to Bogoliubov and Krylov [129], stating that such a measure always exists.

First, let us make some preliminary remarks. As usual, denote by the symbol $C(T)$ the Banach space of all real-valued continuous functions on T. According to the classical Riesz theorem (see, e.g., [59] or [124]), there is a canonical one-to-one correspondence between Radon probability measures on T and linear normalized positive functionals on $C(T)$. In other words, any Radon probability measure μ on T can be canonically identified with some linear normalized positive (hence continuous) functional on $C(T)$. For the sake of convenience, this functional will be denoted by the same symbol μ. Obviously, the set M of all such μ is convex in $(C(T))^*$. Moreover, M turns out to be compact with respect to the weak topology $\sigma((C(T))^*, C(T))$. The latter fact immediately follows from the classical Tychonoff theorem on products of quasicompact spaces, by taking into account the circumstance that M is a closed convex subset of the product

$$\prod_{f \in C(T)} [-||f||, ||f||].$$

In addition, since T is a compact metric space, $C(T)$ is separable and hence M is metrizable, as well. So M is a compact metrizable space, which implies that any sequence of points from M contains at least one convergent subsequence.

Now, take an arbitrary functional $\nu \in M$ and, for each natural number n, define a functional ν_n by the formula

$$\nu_n(f) = \frac{\nu(f) + \nu(f \circ u) + \ldots + \nu(f \circ u^n)}{n+1} \qquad (f \in C(T)).$$

It can easily be seen that $\nu_n \in M$, too. Thus, we obtain the sequence $\{\nu_n : n \in \mathbf{N}\}$ of points from M. Let $\{n(k) : k \in \mathbf{N}\}$ be a strictly increasing family of natural numbers, such that the corresponding partial sequence $\{\nu_{n(k)} : k \in \mathbf{N}\}$ converges to some point $\mu \in M$. Let us verify that

$$\mu(f) = \mu(f \circ u) \qquad (f \in C(T)).$$

Indeed, we may write

$$\mu(f) = lim_{k \to \infty} \nu_{n(k)}(f),$$

$$\mu(f \circ u) = lim_{k \to \infty} \nu_{n(k)}(f \circ u).$$

But a direct calculation shows that

$$\nu_{n(k)}(f) - \nu_{n(k)}(f \circ u) = \frac{\nu(f) - \nu(f \circ u^{n(k)+1})}{n(k)+1},$$

which implies

$$|\nu_{n(k)}(f) - \nu_{n(k)}(f \circ u)| \leq \frac{2||\nu|| \cdot ||f||}{n(k)+1}.$$

Tending k to infinity and taking into account the fact that $n(k) \geq k$, we come to the equality

$$\mu(f) = \mu(f \circ u)$$

for all functions $f \in C(T)$. In addition, since u is a homeomorphism of T onto itself, we have

$$\mu(f \circ u^{-1}) = \mu(f \circ u^{-1} \circ u) = \mu(f)$$

for all $f \in C(T)$. So we can conclude that our measure μ is invariant under the group of transformations, generated by u. However, this group is rather small (in particular, it is at most countable). Markov [147] and Kakutani [105] extended the above result to any commutative group of homeomorphisms of T (their argument is briefly presented in Exercises 14 and 15).

Of course, our survey of invariant and quasi-invariant measures is very far from the complete one. Further information about various properties of such measures can be found in the works mentioned in this section (see also exercises below).

EXERCISES

1. Recall that the standard Borel measure on $\mathbf{R}^n$ is defined as a (unique) measure λ_n satisfying the following conditions:

1) $dom(\lambda_n) = B(\mathbf{R}^n)$;

2) $\lambda_n([0,1]^n) = 1$;

3) λ_n is invariant under some dense subgroup of the group of all translations of $\mathbf{R}^n$.

Deduce directly from conditions 1), 2) and 3) that the standard Borel measure on the space $\mathbf{R}^n$ is invariant under the group of all isometric transformations of $\mathbf{R}^n$ (utilize the fact that the latter group is generated by the family of all symmetries with respect to affine hyperplanes in $\mathbf{R}^n$).

2. Let Γ be a σ-compact locally compact topological group equipped with a Haar measure ν and let ν' denote the completion of ν. Suppose also that G is some separable topological group and let

$$\Psi : \Gamma \to G$$

be a group homomorphism. By applying the Steinhaus property of ν, show that the following assertions are equivalent:

a) Ψ is measurable with respect to ν';

b) Ψ is continuous.

Start with this equivalence and prove (by using the Mackey result formulated in this section and the classical Baire theorem on category) that, for any Polish topological group H, the next two relations are also equivalent:

c) H is locally compact;

d) there exists a Borel probability measure on H quasi-invariant with respect to the group of all left translations of H.

3. Let us recall the theorem of topological measure theory, stating that any σ-finite Borel measure given on a metric space E with non-real-valued measurable weight is concentrated on a closed separable subset of E (see Section 8). By starting with this theorem, establish the following result.

Let G be a metrizable topological group whose weight is not a real-valued measurable cardinal. Suppose also that there exists a nonzero σ-finite Borel measure on G quasi-invariant with respect to a dense (in G) group of left (right) translations of G. Then G is a separable topological group. Consequently, if G is complete, then G is a Polish topological group.

4. Let G be an arbitrary σ-compact locally compact topological group satisfying the inequality

$$card(G) > 2^{\mathbf{c}}$$

where $\mathbf{c}$ denotes the cardinality continuum, and let μ be the left Haar measure on G. Then, by definition, μ is invariant with respect to the group of all left translations of G. Show, on the other hand, that G is not a separable space. In particular, let us suppose that

$$\mathbf{c} = \omega_1, \quad 2^{\mathbf{c}} = \omega_2,$$

and let $\Gamma = \{0,1\}^{\omega_2}$ where $\{0,1\}$ is equipped with the discrete topology. Then Γ is a commutative (with respect to addition modulo 2) nonseparable compact topological group, the weight of Γ is equal to ω_2 and, according to the classical result of Ulam (see e.g. Appendix 3), ω_2 is not a real-valued measurable cardinal. Thus, we see that the assumption of the metrizability of a topological group G is essential for the validity of the result formulated in Exercise 3.

5. Let λ_n denote the standard Borel measure on the n-dimensional Euclidean space $\mathbf{R}^n$ and let μ be an arbitrary nonzero σ-finite measure defined on the Borel σ-algebra of $\mathbf{R}^n$. Demonstrate that these two assertions are equivalent:

a) μ is quasi-invariant with respect to the group of all translations of $\mathbf{R}^n$;

b) there exists a strictly positive Borel function $\phi : \mathbf{R}^n \to \mathbf{R}$ such that

$$\mu(Z) = \int_Z \phi(x)\lambda_n(dx)$$

for each Borel subset Z of $\mathbf{R}^n$.

Prove an analogous result in a more general situation where the space $\mathbf{R}^n$ is replaced by a locally compact Polish topological group G and the standard Borel measure λ_n is replaced by the left Haar measure on G.

6. Give a detailed proof of assertion 2) of Theorem 2.

7. Let G be a subgroup of the additive group $\mathbf{R}^{\mathbf{N}}$, such that

$$G \setminus \mathbf{R}^{(\mathbf{N})} \neq \emptyset.$$

Prove that there exists a family $\{\mu_n : n \in \mathbf{N}\}$ of probability measures, satisfying these two conditions:

a) for each natural number n, the measure μ_n is defined on the Borel σ-algebra of the real line $\mathbf{R}$ and is quasi-invariant with respect to the group of all translations of $\mathbf{R}$;

b) the product measure of the family $\{\mu_n : n \in \mathbf{N}\}$ is not quasi-invariant with respect to G.

8. Check that l_1 is a certain Borel subset of the Polish topological vector space $\mathbf{R}^{\mathbf{N}}$ and show that there exists a family $\{\mu_n : n \in \mathbf{N}\}$ of probability measures, satisfying these two relations:

a) for each natural number n, the measure μ_n is defined on the Borel σ-algebra of $\mathbf{R}$ and is quasi-invariant with respect to the group of all translations of $\mathbf{R}$;

b) for the product measure μ of the family $\{\mu_n : n \in \mathbf{N}\}$, we have $\mu(l_1) > 0$.

Deduce from this fact that there exists a Borel probability measure ν on the Banach space l_1, quasi-invariant with respect to the vector space $\mathbf{R}^{(\mathbf{N})}$ which is everywhere dense in l_1.

9. Let E be a nonseparable normed vector space, let G be the group of all translations of E and let H be a dense subgroup of G. Further, let B be an arbitrary ball in E. Show that, for any σ-finite H-quasi-invariant measure μ on E (not necessarily Borel), the implication

$$B \in dom(\mu) \Rightarrow \mu(B) = 0$$

holds true.

Infer from this fact that there exists no nonzero σ-finite Borel measure on E quasi-invariant with respect to H.

Finally, conclude that, for a Banach space F, the following two assertions are equivalent:

a) F is separable;

b) there exists a Borel probability measure on F quasi-invariant with respect to some everywhere dense vector subspace of F.

10. Let κ be an arbitrary cardinal number. Prove that there exists a Borel probability measure on the topological vector space $\mathbf{R}^{\kappa}$, quasi-invariant with respect to the dense vector subspace $\mathbf{R}^{(\kappa)}$ of $\mathbf{R}^{\kappa}$. Also, check that if κ is strictly greater than the cardinality continuum, then the space $\mathbf{R}^{\kappa}$ is not separable. Deduce from these results that, for a nonmetrizable topological vector space F, assertions a) and b) of the preceding exercise are not, in general, equivalent.

11. Show that the measure μ constructed in Example 4 is metrically transitive with respect to $\mathbf{R}^{(\mathbf{N})}$. In other words, demonstrate that, for any μ-measurable set Z with $\mu(Z) > 0$, there exists a countable family

$$\{g_n \ : \ n \in \mathbf{N}\} \subset \mathbf{R}^{(\mathbf{N})}$$

such that

$$\mu(\mathbf{R}^{\mathbf{N}} \setminus \cup\{g_n(Z) \ : \ n \in \mathbf{N}\}) = 0.$$

12. We recall the notation:

$$l_1 = \{(x_n)_{n\in\mathbf{N}} \in \mathbf{R}^{\mathbf{N}} \ : \ \sum_{n\in\mathbf{N}} |x_n| < +\infty\},$$

$$l_2 = \{(x_n)_{n\in\mathbf{N}} \in \mathbf{R}^{\mathbf{N}} \ : \ \sum_{n\in\mathbf{N}} |x_n|^2 < +\infty\}.$$

Both l_1 and l_2 are Borel vector subspaces of the Polish topological vector space $\mathbf{R}^{\mathbf{N}}$. In addition, l_1 can be considered as a separable Banach space with respect to the natural norm and l_2 can be considered as a separable Hilbert space with respect to the natural inner product.

Prove that there exists a nonzero σ-finite Borel measure ν on the Banach space l_1, invariant with respect to $\mathbf{R}^{(\mathbf{N})}$. Using this result and taking account of the relation

$$l_1 \subset l_2,$$

conclude that, for every infinite-dimensional separable Hilbert space F, there exists a nonzero σ-finite Borel measure on F invariant with respect to some dense vector subspace of F.

13. Let G denote the group of all isometric transformations of $\mathbf{R}^n$, equipped with its natural topology. By applying the classical Lebesgue theorem on density points of Lebesgue measurable sets (see, e.g., [154] or [177]), prove that, for any λ_n-measurable set $Z \subset \mathbf{R}^n$ with $\lambda_n(Z) > 0$, there exists a neighbourhood V of the neutral element of G, such that

$$\lambda_n(g(Z) \cap Z) > 0$$

for all elements $g \in V$; in particular, we have

$$g(Z) \cap Z \neq \emptyset$$

for all $g \in V$. Moreover, using the fact that the standard Borel measure on $\mathbf{R}^n$ is Radon, show that

$$lim_{g \to e} \lambda_n(g(Z) \cap Z) = \lambda_n(Z)$$

where e denotes the neutral element of G.

Formulate and prove an analogous result for a Haar measure on a σ-compact locally compact topological group.

14. Let E be a topological vector space, let K be a nonempty compact convex subset of E and let G be a commutative (with respect to the composition operation) family of affine continuous mappings acting from K into itself. Demonstrate that there exists at least one point of K invariant under G, i.e. there exists a point $x \in K$ satisfying the relation

$$(\forall g \in G)(g(x) = x).$$

This important result is due to Markov and Kakutani.

15. Let T be a nonempty compact topological space and let G be a commutative group of homeomorphisms acting from T into itself. For each element $g \in G$, define the mapping

$$g' : (C(T))^* \to (C(T))^*$$

by the formula

$$g'(\phi) = \phi' \qquad (\phi \in (C(T))^*)$$

where

$$\phi'(f) = \phi(f \circ g) \qquad (f \in C(T)).$$

Check that $G' = \{g' \, : \, g \in G\}$ is a commutative group of continuous affine mappings for which we have

$$g'(M) \subset M \qquad (g' \in G'),$$

where M denotes the compact convex set of all Radon probability measures on T. By applying to G' and to M the result of Exercise 14, show that there exists a Radon probability measure μ on T invariant under G, i.e.

$$\mu(g(X)) = \mu(X) \qquad (g \in G, \ X \in B(T)).$$

16. Let G be a commutative compact topological group. By using the result of Exercise 15, demonstrate the existence of a Haar probability measure on G. In addition, by applying the classical Fubini theorem to the product space $G \times G$, prove the uniqueness of a Haar probability measure on G.

17. Let S be a sphere in $\mathbf{R}^n$ $(n \geq 2)$ equipped with the $(n-1)$-dimensional Lebesgue probability measure μ which is invariant under the group G of all rotations of S about its centre. Fix a natural number $k > 0$. Let A be an arbitrary μ-measurable subset of S with $\mu(A) < 1/k$ and let B be an arbitrary subset of S with $card(B) \leq k$. Show that there exists at least one rotation $g \in G$ for which we have

$$g(A) \cap B = \emptyset.$$

10. Anderson inequality and unimodal distributions

Let F be a topological vector space and let μ be a probability measure on $(F, B(F))$.

Definition 1. We say that μ is unimodal in the sense of Anderson (or that μ satisfies the Anderson inequality) if, for each Borel convex symmetric (with respect to zero) set $D \subset F$, the following relation is fulfilled:

$$(\forall y \in F)(\forall t \in [0,1])(\mu(D + ty) \geq \mu(D + y)).$$

In particular, putting $t = 0$, we have

$$(\forall y \in F)(\mu(D) \geq \mu(D + y)).$$

Thus, in our further considerations we understand the unimodality as an unimodality at zero, and it means that the μ-measure of a Borel convex symmetric set D is a decreasing function when D moves on a fixed half-line passing through the origin (the movement of D starts from the origin, of course). This geometric property of μ turns out to be rather important in various applications, so simple criteria for recognizing it are often needed. The integral inequality of Anderson (see Theorem 5 from Section 2) leads to some sufficient conditions under which a given probability measure μ on $\mathbf{R}^n$ is unimodal. Namely, we recall that μ turns out to be unimodal if it is absolutely continuous with respect to the standard Lebesgue measure on $\mathbf{R}^n$ and the Radon-Nikodym density f of μ satisfies the conditions:

1) $f(-x) = f(x)$ for all $x \in \mathbf{R}^n$;

2) for any $\theta > 0$, the set

$$\{x \in \mathbf{R}^n \ : \ f(x) \geq \theta\}$$

is convex in $\mathbf{R}^n$.

The above-mentioned conditions are sometimes very helpful and, in general, are not hard for verification.

Example 1. Let the density f of a probability distribution μ in $\mathbf{R}^n$ has the form

$$f(x) = \beta \cdot exp(-||Ax||^{\alpha}) \qquad (x \in \mathbf{R}^n)$$

where $||\cdot||$ denotes the usual norm in $\mathbf{R}^n$, a real number α is strictly positive, A is a linear mapping (operator) acting from $\mathbf{R}^n$ into $\mathbf{R}^n$, and β is a normalized constant. Then it can easily be checked that conditions 1) and 2) hold true for f. Consequently, μ turns out to be an unimodal distribution.

Example 2. Let the density f of a probability distribution in the space $\mathbf{R}^n$ be of the form:

$$f(x) = f_1(x_1)f_2(x_2)...f_n(x_n) \qquad (x = (x_i)_{1 \leq i \leq n} \in \mathbf{R}^n),$$

where $(f_i)_{1 \leq i \leq n}$ are symmetric logarithmically upper convex real-valued functions. Then the Anderson inequality implies at once that f is the density of some symmetric unimodal distribution in $\mathbf{R}^n$.

Example 3. Let the density f of a probability distribution in the plane $\mathbf{R}^2$ be of the form:

$$f(x) = \beta \cdot exp(-(|x_1|^{1/2} + |x_2|^{1/2})) \qquad (x = (x_1, x_2) \in \mathbf{R}^2).$$

In this case, condition 1) is obviously fulfilled. However, for any $\theta > 0$, the set

$$\{x \in \mathbf{R}^2 \ : \ f(x) \geq \theta\}$$

is not convex, which can easily be observed. Thus, here we are not able to utilize the criterion mentioned above. Nevertheless, it will be demonstrated below, by applying an essentially different argument, that f turns out to be the density of a unimodal distribution in $\mathbf{R}^2$.

The last example shows us that, in some natural situations, the criterion of unimodality based on the Anderson integral inequality does not work. Besides, if we wish to formulate similar criteria for probability distributions in various infinite-dimensional vector spaces, we immediately meet difficulties connected with the nonexistence of an appropriate analogue of the classical Lebesgue measure for such spaces. In order to avoid the difficulties of this sort, we determine below a certain class of unimodal distributions in $\mathbf{R}^n$, containing distributions similar to the one described in Example 3 and many others. Also, we shall determine a suitable class of unimodal distributions for the infinite-dimensional topological vector space $\mathbf{R}^{\mathbf{N}}$ and investigate some properties of that class.

However, before starting the investigation of unimodal distributions in general topological vector spaces F, it is reasonable to examine the most simple case when $F = \mathbf{R}$. We recall that, in the theory of probability distributions, a certain class of distributions on the real line $\mathbf{R}$ is studied in detail and extensively. We mean here the class of distributions which are unimodal in the sense of Khinchin (see [119]). Let us give a precise definition of this class (in the form convenient for our further considerations). As usual, we denote by λ_1 the standard one-dimensional Lebesgue measure on the real line $\mathbf{R}$.

Definition 2. A probability measure μ on $(\mathbf{R}, B(\mathbf{R}))$ is unimodal (in the sense of Khinchin) with a mode at a point $a \in \mathbf{R}$ if:

1) μ is locally absolutely continuous with respect to λ_1 at all points of the set $\mathbf{R} \setminus \{a\}$;

2) the density of μ is increasing almost everywhere on the open interval $]-\infty, a[$ and is decreasing almost everywhere on the open interval $]a, +\infty[$.

We recall that a Borel measure ν on a topological vector space F is symmetric (with respect to 0) if

$$\nu(D) = \nu(-D)$$

for all sets $D \in B(F)$.

It can easily be observed that if a Khinchin unimodal measure μ on $\mathbf{R}$ is symmetric, then the point $a = 0$ is its mode and the density f of the absolutely continuous component of μ satisfies the relation

$$f(x) = f(-x)$$

for almost all $x \in \mathbf{R}$. In view of this evident fact, a natural question arises concerning the compatibility of the two concepts of unimodality: in the sense of Khinchin and in the sense of Anderson. In order to answer this question, we first formulate the following auxiliary statement.

Lemma 1. *Any symmetric probability measure on* $\mathbf{R}$ *unimodal in the sense of Khinchin is also unimodal in the sense of Anderson.*

The proof of this lemma is not hard and we leave it to the reader. We only wish to stress the circumstance that the assumption on μ (i.e. the symmetricity of μ) is essential in the formulation of the lemma and cannot be dropped. The converse assertion is much more difficult to prove but, nevertheless, it holds true without any additional assumptions.

Lemma 2. *Each probability measure μ on $\mathbf{R}$ unimodal in the Anderson sense is unimodal in the Khinchin sense and one of the modes of μ coincides with 0.*

Proof. We shall demonstrate even a more general fact showing that the unimodality in the sense of Anderson can be weakened here. Namely, we only suppose that, for any $t \geq 0$, the following two relations are satisfied:

$$(1) \quad \mu([a,b]) \geq \mu([a,b]+t) \quad (a \geq 0,\ b \geq 0,\ b \geq a);$$

$$(2) \quad \mu([a,b]) \geq \mu([a,b]-t) \quad (a \leq 0,\ b \leq 0,\ b \geq a).$$

Obviously, these relations are implied by the Anderson inequality.

Also, (1) and (2) immediately yield that our measure μ can possess an atom only at the point 0, i.e. we have

$$(3) \quad \mu(\{x\}) = 0 \quad (x \in \mathbf{R} \setminus \{0\}).$$

Now, let us establish that μ is absolutely continuous with respect to the Lebesgue measure λ_1 on the interval $]0,+\infty[$. For this purpose, it is sufficient to show that, for each segment $[c,d] \subset]0,+\infty[$, the relation

$$(4) \quad sup_{[\alpha,\beta]\subset[c,d],\alpha<\beta} \frac{\mu([\alpha,\beta])}{\beta-\alpha} < +\infty$$

holds true. Suppose otherwise, i.e. there exists a segment $[c,d]$ contained in $]0,+\infty[$ such that, for any $\delta > 0$, a segment $[\alpha,\beta] \subset [c,d]$ can be found for which

$$(5) \quad 0 < \beta - \alpha < \delta, \quad \frac{\mu([\alpha,\beta])}{\beta-\alpha} \geq 1/\delta.$$

Let us put

$$\gamma = \beta - \alpha, \quad k = [c/\gamma],$$

where the symbol $[x]$ denotes, as usual, the largest natural number less than or equal to $x \in \mathbf{R}$. Let $\delta \leq c/3$. It is clear that

$$[0,\gamma] \cup [\gamma, 2\gamma] \cup \ldots \cup [(k-1)\gamma, k\gamma] \subset [0,c].$$

Taking into account (1), (3) and (5), we infer from this inclusion that

$$1 \geq \mu([0,c]) \geq \sum_{1\leq i\leq k} \mu([(i-1)\gamma, i\gamma]) \geq k\mu([\alpha,\beta]) \geq$$

$$(\frac{c}{\beta-\alpha} - 1)\mu([\alpha,\beta]) \geq 2,$$

which is impossible. The contradiction obtained shows us that relation (4) holds and, in particular, μ is absolutely continuous with respect to λ_1 on $]0, +\infty[$. Applying a similar argument, we are able to establish that μ is absolutely continuous with respect to λ_1 on $]-\infty, 0[$ (here (2) must be utilized instead of (1)). Summarizing, we conclude that μ turns out to be locally absolutely continuous with respect to λ_1 at all points of $\mathbf{R} \setminus \{0\}$.

Now, we are going to demonstrate that the density f of the absolutely continuous component of μ is increasing almost everywhere on the interval $]-\infty, 0[$ and is decreasing almost everywhere on the interval $]0, +\infty[$. For this purpose, we denote by $D(f)$ the set of all those points $t \in \mathbf{R}$ for which

$$f(t) = lim_{h \to 0, h \neq 0} \, (1/h) \int_t^{t+h} f(x)dx.$$

Since the function f is locally integrable on $\mathbf{R} \setminus \{0\}$, almost all (with respect to λ_1) points of the real line are in the set $D(f)$. Further, choose any two points s and t from the set $D(f) \cap \,]0, +\infty[$ and assume that $s < t$. In virtue of relation (1), we may write

$$\int_s^{s+h} f(x)dx = \mu([s, s+h]) \geq \mu([t, t+h]) = \int_t^{t+h} f(x)dx$$

for any $h > 0$. Consequently,

$$f(s) = lim_{h \to 0, h \neq 0} \, (1/h) \int_s^{s+h} f(x)dx \geq$$

$$lim_{h \to 0, h \neq 0} \, (1/h) \int_t^{t+h} f(x)dx = f(t).$$

In a similar way, we can establish that $f(s) \leq f(t)$ for any two points s and t belonging to the set $D(f) \cap \,]-\infty, 0[$ and satisfying the relation $s < t$. This completes the proof of Lemma 2.

From the results presented above, we obtain at once the following statement.

Lemma 3. *Among the symmetric probability measures on* $\mathbf{R}$, *the class of all unimodal measures in the sense of Anderson coincides with the class of all unimodal measures in the sense of Khinchin.*

Thus, we can simply speak of unimodal symmetric probability distributions on the real line, without specifying the type of unimodality (in the

Anderson sense or in the Khinchin sense). But, in general, the unimodality in the Anderson sense seems to be more preferable because it admits further generalizations to topological vector spaces of arbitrary dimension. However, the class of unimodal distributions in the Khinchin sense is important from the probabilistic point of view. We wish to mention here one interesting probabilistic characterization of distributions from this class.

Example 4. Consider any two independent random variables V and Z on a probability space (Ω, S, P), satisfying the properties described below. Namely, we shall assume that V is uniformly distributed on the unit segment $[0,1]$ (i.e. the distribution of V coincides with the standard Lebesgue measure on $[0,1]$) and Z has a discrete distribution of the type

$$P(\{Z = nh\}) = q_n \qquad (n \geq 1)$$

where $h > 0$ is a fixed real number and $(q_n)_{n \geq 1}$ is some sequence of positive reals. The distribution μ of the product VZ can easily be found and turns out to be of the form

$$\mu(C) = P(\{VZ \in C\}) = \sum_{n \geq 1} P(\{V \in \frac{C}{nh}\})q_n$$

where C is an arbitrary Borel subset of $\mathbf{R}$. The measure μ is absolutely continuous with respect to λ_1 and its density q can be written in the form

$$(1) \quad q(t) = \sum_{n \geq 1} (\frac{q_n}{nh}) I_{[0,1[}(\frac{t}{nh}) \qquad (t \in \mathbf{R})$$

where I_C denotes, as usual, the indicator of a set $C \subset \mathbf{R}$.

Let us introduce a step function

$$p : \mathbf{R} \rightarrow \mathbf{R}$$

by the equality

$$p(t) = \sum_{n \geq 1} p_n I_{[(n-1)h, nh[}(t) \qquad (t \in \mathbf{R})$$

where $(p_n)_{n \geq 1}$ is any decreasing sequence of positive real numbers. Note that p is the density of a Khinchin unimodal distribution (concentrated on $[0, +\infty[$) if and only if

$$h \sum_{n \geq 1} p_n = 1.$$

Using the trivial fact that

$$p(t) = (p_1 - p_2)I_{[0,h[}(t) + (p_2 - p_3)(I_{[0,h[}(t) + I_{[h,2h[}(t)) + \dots,$$

we may express the function p in the form

$$(2) \quad p(t) = \sum_{n\geq 1} \frac{q'_n}{nh} I_{[0,1[}\left(\frac{t}{nh}\right) \qquad (t \in \mathbf{R}),$$

where

$$q'_n = (p_n - p_{n+1})nh \qquad (n \geq 1),$$

and, in particular, we get

$$\sum_{n\geq 1} q'_n = h \sum_{n\geq 1} p_n.$$

Comparing (1) and (2), we conclude that p is the density of a Khinchin unimodal distribution (with a mode at zero) if and only if p is the density of the distribution of some product of two independent random variables, one of which is uniformly distributed on the segment $[0, 1]$.

This argument can easily be extended to any Khinchin unimodal distribution on $\mathbf{R}$ whose mode coincides with zero and for which the density is a step function with discontinuity points $(\pm nh)_{n\in\mathbf{N}}$. Tending a parameter h to 0, we can approximate any Khinchin unimodal distribution (with mode at zero) by unimodal distributions described above. This approach (due to Shepp) leads to the following statement.

Lemma 4. *A probability measure μ on $(\mathbf{R}, B(\mathbf{R}))$ is Khinchin unimodal (with mode 0) if and only if μ coincides with the distribution of a product of two independent random variables V and Z, such that V is uniformly distributed on $[0, 1]$. In other words,*

$$\mu(C) = P(\{VZ \in C\}) = \int_0^1 \mu_1(C/v)dv \qquad (C \in B(\mathbf{R})),$$

where μ_1 denotes the distribution of Z.

Note, in addition, that the preceding lemma may be regarded as a probabilistic interpretation of a widely known statement of Khinchin and was originally proved by Zolotarev who utilized purely analytic methods (in this connection, see [74] and [201]).

Starting with Lemmas 3 and 4, we are able to establish the next auxiliary proposition which will be essentially applied in our further considerations.

Lemma 5. *A symmetric probability measure μ on $\mathbf{R}$ is unimodal if and only if μ is the distribution of a product of two independent random variables V and Z, such that V is uniformly distributed on the segment $[-1, 1]$; in other words,*

$$\mu(C) = (1/2) \int_{-1}^{1} \mu_1(C/v)dv \qquad (C \in B(\mathbf{R})),$$

where μ_1 denotes the distribution of Z.

Proof. In accordance with Lemmas 3 and 4, the measure μ can be represented in the form

$$\mu(C) = \int_{0}^{1} \mu_1(C/v)dv \qquad (C \in B(\mathbf{R})).$$

But, since μ is symmetric, we may write

$$\mu(C) = (1/2)\mu(C) + (1/2)\mu(-C) =$$

$$(1/2) \int_{0}^{1} \mu_1(C/v)dv + (1/2) \int_{0}^{1} \mu_1(-C/v)dv = (1/2) \int_{-1}^{1} \mu_1(C/v)dv$$

for each Borel subset C of $\mathbf{R}$. This completes the proof of the lemma.

Recall that a random element X with values in a topological vector space F (i.e. F-valued random element) is said to be symmetric if the distribution of X is a symmetric probability measure. In addition, we shall say that a random element X is unimodal if the distribution of X is a unimodal probability measure (in the sense of Anderson).

Now, we fix a finite-dimensional Euclidean space F and describe one important class of unimodal symmetric random vectors in this space.

Theorem 1. *Let X be a random vector in the space $F = \mathbf{R}^m$, where $m \geq 1$, and suppose that the coordinates of X are mutually independent unimodal symmetric random variables. Let*

$$A : F \to F$$

be an arbitrary linear mapping. Then the random vector

$$Y = A \circ X = A(X) = AX$$

is unimodal and symmetric, too.

Proof. The fact that Y is symmetric follows directly from the assumptions (because X is symmetric and A is linear). The main difficulty lies in the proof of the unimodality of Y. We shall establish the latter fact gradually, by considering several cases.

First, let us remark that we may replace X by any its copy and then to apply an argument to such a copy. In virtue of Lemma 5, X admits a representation in the form

$$X = (V_1 Z_1, V_2 Z_2, ..., V_m Z_m)$$

where

$$V_1, V_2, \ldots, V_m, Z_1, Z_2, \ldots, Z_m$$

are mutually independent random variables and, in addition, each of the random variables $V_1, V_2, ..., V_m$ is uniformly distributed on the segment $[-1, 1]$. The corresponding basic probability space (Ω, S, P) on which the above-mentioned family of random variables is defined can easily be described. Indeed, it suffices to put

$$(\Omega, S, P) = (\mathbf{R}^{2m}, B(\mathbf{R}^{2m}), (\prod_{1 \le j \le m} \lambda'_j) \times (\prod_{1 \le j \le m} \nu_j))$$

where, for each natural index $j \in [1, m]$, the symbol λ'_j denotes the distribution of V_j and the symbol ν_j denotes the distribution of Z_j. It is also convenient to put

$$\nu = \nu_1 \times \nu_2 \times ... \times \nu_m.$$

So ν turns out to be the distribution of the random vector

$$Z = (Z_1, Z_2, ..., Z_m).$$

Now, in order to prove the unimodality of Y, we are going to demonstrate that, for any compact strongly convex symmetric (with respect to zero) body $D \subset F$, the inequality

$$(1) \quad P(\{Y \in D + ty\}) \ge P(\{Y \in D + y\})$$

is satisfied whenever $0 \le t \le 1$ and $y \in F$. Note that the validity of (1) for compact strongly convex symmetric bodies in F implies its validity for arbitrary Borel convex symmetric sets in F (see Exercise 2).

To show that (1) holds true, consider the following two cases.

1. $A = I$ where I is the identity transformation of F. In this case $Y = X$. Denote by J the family of all subsets of $\{1, 2, ..., m\}$ (i.e. J is the

power set of $\{1, 2, ..., m\}$). Further, for any set $\{j_1, j_2, ..., j_k\}$ from J, define the set $G_{j_1,j_2,...,j_k}$ consisting of all those vectors $z \in F$ for which

$$(\forall j)(j \in \{j_1, j_2, ..., j_k\} \Rightarrow z_j \neq 0),$$

$$(\forall j)(j \notin \{j_1, j_2, ..., j_k\} \Rightarrow z_j = 0).$$

Obviously, the family of sets

$$(G_{j_1,j_2,...,j_k})_{\{j_1,j_2,...,j_k\} \in J}$$

forms a finite partition of the space F. For any set $\{j_1, j_2, ..., j_k\}$ from J, we also define the vector space

$$F_{j_1,j_2,...,j_k} \subset F$$

consisting of all those elements $x \in F$ which satisfy the condition

$$(\forall j)(j \notin \{j_1, j_2, ..., j_k\} \Rightarrow x_j = 0).$$

In addition, for each $z \in F$, we put

$$C_{j_1,j_2,...,j_k}(z) = \{x \in F_{j_1,j_2,...,j_k} \ : \ |x_{j_1}| < |z_{j_1}|, \ . \ . \ . \ , \ |x_{j_k}| < |z_{j_k}|\}.$$

Notice that if $z \in G_{j_1,j_2,...,j_k}$, then the random vector

$$(z_1 V_1, z_2 V_2, ..., z_m V_m)$$

takes its values in the space $F_{j_1,j_2,...,j_k}$ and, under the assumption

$$\{j_1, j_2, ..., j_k\} \neq \emptyset,$$

this vector is uniformly distributed in $C_{j_1,j_2,...,j_k}(z)$. Consequently, the relations

$$\{j_1, j_2, ..., j_k\} \neq \emptyset, \quad z \in G_{j_1,j_2,...,j_k}, \quad C \in B(F)$$

imply the equality

$$P(\{(z_1 V_1, z_2 V_2, ..., z_m V_m) \in C\}) =$$

$$\frac{\mu_{j_1,j_2,...,j_k}(C \cap C_{j_1,j_2,...,j_k}(z))}{|z_{j_1}| \cdot |z_{j_2}| \cdot ... \cdot |z_{j_k}|}$$

where $\mu_{j_1,j_2,...,j_k}$ denotes the standard Lebesgue measure on the Euclidean space $F_{j_1,j_2,...,j_k}$. For $G_\emptyset = \{0\}$, we put

$$P(\{(z_1 V_1, z_2 V_2, ..., z_m V_m) \in C\}) = I_C(0) \qquad (z = 0)$$

where I_C is the indicator of C. Summarising all said above, we come to the formula

$$(2) \quad P(\{X \in C\}) = \nu(\{0\})I_C(0) +$$

$$\sum_{\{j_1,j_2,\ldots,j_k\} \in J \setminus \{\emptyset\}} \int_{G_{j_1,j_2,\ldots,j_k}} \frac{\mu_{j_1,j_2,\ldots,j_k}(C \cap C_{j_1,j_2,\ldots,j_k}(z))}{|z_{j_1}| \cdot |z_{j_2}| \cdot \ldots \cdot |z_{j_k}|} \nu(dz).$$

Note that if all m distributions of the components of the random vector Z are absolutely continuous with respect to the standard Lebesgue measure on $\mathbf{R}$, then (2) has an essentially simplified form. Namely, in this case, (2) contains only one summand corresponding to the whole set

$$\{j_1, j_2, \ldots, j_k\} = \{1, 2, \ldots, m\}.$$

Now, if $\{j_1, j_2, \ldots, j_k\} \neq \emptyset$ and

$$t \in [0,1], \quad y \in F, \quad z \in G_{j_1,j_2,\ldots,j_k},$$

then, for every convex symmetric body $D \subset F$, we have the inequality

$$\mu_{j_1,j_2,\ldots,j_k}((D + ty) \cap C_{j_1,j_2,\ldots,j_k}(z)) \geq$$

$$\mu_{j_1,j_2,\ldots,j_k}((D + y) \cap C_{j_1,j_2,\ldots,j_k}(z))$$

which, in fact, was established by the argument used in the proof of Theorem 5 from Section 2. Besides, it is clear that

$$I_{D+ty}(0) \geq I_{D+y}(0).$$

Hence the relation (1) holds true, and the first step of our proof has been done.

2. Consider now the case of an arbitrary linear mapping A acting from F into F. Fix an element $y \in F$ and a real number $t \in [0,1]$. Suppose first that

$$y \in ran(A) = \{A(x) \ : \ x \in F\}.$$

Then there exists an element $x \in F$ such that $y = A(x)$. Applying the result of case 1, we may write

$$P(\{Y \in D + ty\}) = P(\{X \in A^{-1}(D + ty)\}) =$$

$$P(\{X \in A^{-1}(D) + tx\}) \geq P(\{X \in A^{-1}(D) + x\}) =$$

$$P(\{X \in A^{-1}(D + y)\}) = P(\{Y \in D + y\}).$$

Suppose now that $y \in F \setminus ran(A)$. For a strictly positive real number ε, consider the linear mapping

$$A_\varepsilon = A + \varepsilon I$$

where I denotes again the identity transformation of F, and introduce the random vector

$$Y_\varepsilon = A_\varepsilon \circ X = A_\varepsilon(X).$$

If ε is sufficiently small, then A_ε turns out to be a surjection onto F, so, according to the preceding result, we get

$$(3) \quad P(\{Y_\varepsilon \in D + ty\}) \geq P(\{Y_\varepsilon \in D + y\})$$

for every strongly convex symmetric body $D \subset F$. Assume that

$$(4) \quad P(\{Y \in bd(D + ty)\}) = 0, \quad P(\{Y \in bd(D + y)\}) = 0,$$

where the symbol $bd(\cdot)$ denotes, as usual, the boundary operator. Then, tending ε to zero, we see that the random vector Y_ε converges to the random vector Y, and hence the inequality (3) yields us the required inequality (1). Taking into account the strong convexity of D and expression (2) for the distribution of our original random vector X, we see that (4) can fail to be true only in the case when at least one of the relations

$$0 \in bd(D + ty), \quad 0 \in bd(D + y)$$

is fulfilled. But, in such a case, we replace the body D by the set

$$D_n = (1 + (1/n))D \quad (n \in \mathbf{N}, n \geq 1).$$

For sufficiently large n, the set D_n satisfies the corresponding analogue of (4) whence it follows that

$$P(\{Y \in D_n + ty\}) \geq P(\{Y \in D_n + y\}).$$

It remains to point out that

$$D + ty = \cap_{n \geq 1}(D_n + ty),$$

$$D + y = \cap_{n \geq 1}(D_n + y)$$

and to tend n to infinity in the last inequality.

Thus, in all possible situations, (1) is valid, which completes the proof of our theorem.

Actually, Theorem 1 states that any probability measure representable as an image, under a linear mapping, of a product of finitely many symmetric unimodal distributions on $\mathbf{R}$, is a symmetric unimodal measure in the corresponding Euclidean space. We can directly apply the result obtained to Example 3 from this section. So we may assert that the function f described in Example 3 is the density of some symmetric unimodal measure on the plane $\mathbf{R}^2$.

Also, Theorem 1 enables us to define a certain class of Borel probability measures in the infinite-dimensional topological vector space $\mathbf{R}^{\mathbf{N}}$ and to establish a suitable analogue of this theorem for that class.

Definition 3. We say that a probability measure μ given on the space $(\mathbf{R}^{\mathbf{N}}, B(\mathbf{R}^{\mathbf{N}}))$ belongs to the class $M' = M'(\mathbf{R}^{\mathbf{N}})$ if there exist a countable family $(\mu_k)_{k\in\mathbf{N}}$ of symmetric unimodal measures on $(\mathbf{R}, B(\mathbf{R}))$ and an infinite triangular matrix $A = (a_{nk})_{n\in\mathbf{N},k\in\mathbf{N}}$ (as in Example 9 of Section 6), such that

$$\mu = (\prod_{k\in\mathbf{N}} \mu_k) \circ A^{-1}$$

where A is the linear mapping of $\mathbf{R}^{\mathbf{N}}$ into itself, associated with the above-mentioned matrix.

We shall denote such a measure $\mu \in M'$ by the symbol

$$\mu = \mu((\mu_k)_{k\in\mathbf{N}}, A).$$

It is clear that μ may be regarded as the distribution in $\mathbf{R}^{\mathbf{N}}$ of the random element $Y = (Y_k)_{k\in\mathbf{N}}$ defined by

$$Y = A \circ X = A(X)$$

where $X = (X_k)_{k\in\mathbf{N}}$ is a sequence of independent random variables and, for each $k \in \mathbf{N}$, the distribution of X_k coincides with μ_k.

Theorem 2. *Every measure μ from the class M' is a symmetric unimodal measure on $(\mathbf{R}^{\mathbf{N}}, B(\mathbf{R}^{\mathbf{N}}))$.*

Proof. Let $\mu \in M'$. For each $m \in \mathbf{N}$, consider the canonical projection

$$Pr_m : \mathbf{R}^{\mathbf{N}} \to \mathbf{R}^{m+1}$$

given by the formula

$$Pr_m(x) = (x_0, ..., x_m) \qquad (x = (x_k)_{k\in\mathbf{N}} \in \mathbf{R}^{\mathbf{N}}).$$

In accordance with Definition 3 and the result of Theorem 1, the measure μ_m determined by the equality

$$\mu_m = \mu \circ Pr_m^{-1}$$

is symmetric and unimodal in $\mathbf{R}^{m+1}$. Consequently, for any convex symmetric (with respect to zero) body D in $\mathbf{R}^{\mathbf{N}}$, for all $t \in [0,1]$ and for each $y \in \mathbf{R}^{\mathbf{N}}$, we may write

$$\mu(D + ty) = lim_{m\to\infty}\mu_m(Pr_m(D + ty)) =$$

$$lim_{m\to\infty}\mu_m(Pr_m(D) + tPr_m(y)) \geq lim_{m\to\infty}\mu_m(Pr_m(D) + Pr_m(y)) =$$

$$lim_{m\to\infty}\mu_m(Pr_m(D + y)) = \mu(D + y),$$

which shows us that μ is unimodal. Evidently, μ is also symmetric, so Theorem 2 has been proved.

Example 5. Let $X = (X_k)_{k\in\mathbf{N}}$ be a sequence of independent random variables with the corresponding distribution densities

$$p_k(t) = d_k exp(-b_k|t|^{a_k}) \qquad (t \in \mathbf{R},\ k \in \mathbf{N})$$

where

$$\{a_k \ :\ k \in \mathbf{N}\} \subset]0, +\infty[, \quad \{b_k \ :\ k \in \mathbf{N}\} \subset]0, +\infty[$$

and $(d_k)_{k\in\mathbf{N}}$ is the sequence of normalized coefficients. Let

$$A = (a_{kj})_{k\in\mathbf{N}, j\in\mathbf{N}}$$

be an arbitrary infinite triangular matrix. Then, in virtue of Theorem 2, the random sequence $Y = (Y_k)_{k\in\mathbf{N}}$ defined by the formula

$$Y_k = \sum_{0\leq j\leq k} a_{kj}X_j \qquad (k \in \mathbf{N})$$

turns out to be a symmetric and unimodal random element in the space $\mathbf{R}^{\mathbf{N}}$.

One subclass of $M' = M'(\mathbf{R}^{\mathbf{N}})$ is very important from the probabilistic viewpoint. This subclass is constituted by the so-called Gaussian measures in $\mathbf{R}^{\mathbf{N}}$. We wish to recall here some basic facts concerning Gaussian measures and Gaussian random elements. These facts will be applied in the next sections of our book.

As usual, we suppose that all random elements under consideration are defined on a fixed probability space (Ω, S, P). We also recall that if X is a random variable on Ω, then the symbol $E(X)$ denotes the mean value (or the mathematical expectation) of X, i.e.

$$E(X) = \int_{\Omega} X(\omega)P(d\omega)$$

assuming that this integral exists. The mean value of X can be expressed in terms of the distribution μ_X. Namely, the following simple formula is true

$$E(X) = \int_{\mathbf{R}} x\mu_X(dx)$$

which is a particular case of a much more general formula (see Exercise 12).

A random variable X is called Gaussian (or normally distributed), with parameters a and σ^2 where

$$a \in \mathbf{R}, \quad \sigma^2 \in [0, +\infty[,$$

if its characteristic functional ϕ_X (see Section 7) admits the representation

$$\phi_X(t) = E(exp(itX)) = exp(ita - \frac{t^2\sigma^2}{2}) \qquad (t \in \mathbf{R}).$$

Here i denotes the imaginary number for which $i^2 = -1$. We shall also say, in the sequel, that a Gaussian random variable X with parameters a and σ^2 is $N(a, \sigma^2)$-distributed. It can easily be checked that, for such a variable, we have

$$E(X) = a, \quad E((X - a)^2) = \sigma^2.$$

In other words, a turns out to be the mathematical expectation of X and σ^2 coincides with the variance of X.

Now, let F be some topological vector space and let H be a subset of F^* separating the points of F. A random element X in $(F, B(F, H))$ (i.e. an H-weak random element in F) is said to be Gaussian if, for each linear functional $h \in lin(H)$, the random variable $h \circ X$ turns out to be Gaussian. If, in addition to this, we have

$$a_h = E(h \circ X) = 0$$

for all $h \in lin(H)$, then the given Gaussian random element X is called centered.

In connection with the definition above, let us make some remarks about the terminology.

A family $(X_t)_{t\in T}$ is said to be a Gaussian family of random variables (or a family of jointly Gaussian random variables) if $(X_t)_{t\in T}$ is a Gaussian random element in the space

$$(\mathbf{R}^T, B(\mathbf{R}^T, (pr_t)_{t\in T})).$$

If T is finite, then we are concerned with a Gaussian vector, if T is infinite and countable, then we speak of a Gaussian sequence, and, for an arbitrary T, we speak of a Gaussian stochastic process.

In other words, a family $(X_t)_{t\in T}$ of random variables turns out to be Gaussian if and only if any linear combination of each finite subfamily of $(X_t)_{t\in T}$ is a Gaussian random variable. In particular, we see that if a given family $(X_t)_{t\in T}$ is Gaussian, then all subfamilies of

$$lin((X_t)_{t\in T})$$

are Gaussian, too.

Further, let F_1 and F_2 be two topological vector spaces, let

$$H_1 \subset F_1^*, \qquad H_2 \subset F_2^*$$

be any two sets separating the points of F_1 and F_2 respectively, and let X_1 and X_2 be any two Gaussian random elements in the spaces $(F_1, B(F_1, H_1))$ and $(F_2, B(F_2, H_2))$ respectively. We say that the pair (X_1, X_2) is jointly Gaussian if it can be regarded as a Gaussian random element in the product space

$$(F_1, B(F_1, H_1)) \times (F_2, B(F_2, H_2)).$$

It immediately follows from this definition that the pair (X_1, X_2) will be jointly Gaussian if and only if, for each element $h_1 \in lin(H_1)$ and for each element $h_2 \in lin(H_2)$, the pair

$$(h_1 \circ X_1, h_2 \circ X_2)$$

is a Gaussian random vector in the plane $\mathbf{R}^2$.

Example 6. Any family $(X_t)_{t\in T}$ of mutually independent Gaussian random variables forms a jointly Gaussian family. If, in addition, all X_t have the parameters $a = 0$ and $\sigma^2 = 1$, then $(X_t)_{t\in T}$ is called to be standard.

We shall see below that, in some sense, the standard Gaussian families are basic for all other Gaussian families.

Let F be a topological vector space and let $H \subset F^*$ separate the points of F. For any random element X whose values belong to $(F, B(F, H))$ and for which the relations

$$E((h \circ X)^2) < +\infty \qquad (h \in H)$$

are satisfied, the bilinear form

$$Q(h, g) = E((h \circ X - a_h)(g \circ X - a_g)) \qquad (h \in lin(H),\ g \in lin(H))$$

is defined. As usual, Q is called the covariance form of X. Obviously, Q is positive, i.e.

$$Q(h, h) \geq 0 \qquad (h \in lin(H)).$$

Further, Q is symmetric, i.e.

$$Q(h, g) = Q(g, h) \qquad (h \in lin(H),\ g \in lin(H)).$$

Note also that Q is uniquely determined by the corresponding function K defined on $H \times H$ by

$$K(h, g) = Q(h, g) \qquad (h \in H,\ g \in H).$$

The latter function is usually called the covariance function of X (or, simply, the covariance of X) and is denoted by the symbol $cov(X)$. It should be noted that the function K is positive, i.e. for each finite sequence $(a_1, a_2, ..., a_m)$ of real numbers and for each finite sequence $(h_1, h_2, ..., h_m)$ of elements from H, we have

$$\sum_{1 \leq n \leq m,\ 1 \leq j \leq m} a_n a_j K(h_n, h_j) \geq 0,$$

and, simultaneously, K is symmetric, i.e.

$$K(h, g) = K(g, h) \qquad (h \in H,\ g \in H).$$

Obviously, if H is a vector subspace of F^*, then the covariance form Q and the covariance function K coincide. But it is frequently convenient to distinguish these two notions because, in many practical situations, the covariance function K possesses rather simple expression.

Example 7. Let $(X_1, X_2, ..., X_m)$ be a centered Gaussian random vector in the space $\mathbf{R}^m$. Then we have

$$cov(X) = (E(X_k X_j))_{(k,j) \in [1,m]^2}.$$

In such a case, the covariance of X may be regarded as an $(m \times m)$-matrix which is also called the covariance matrix of X. Let us remark that if the components of X are mutually independent, then

$$cov(X) = I$$

where I is the unit $(m \times m)$-matrix (i.e. all elements of I on the main diagonal are equal to 1 and all other elements of I are zero). An analogous interpretation in terms of matrices can be applied to covariance functions of various random sequences.

In our further considerations, we are forced to utilize the standard techniques from the theory of matrices. In this connection, let us adopt the following principle concerning rows and columns of finite matrices. Namely, the notation

$$A = (a_{kj})_{k \in [1,n],\ j \in [1,m]}$$

means that the matrix A has n rows and m columns, and a_{kj} is the element obtained by intersecting the k-th row with the j-th column. This convention implies naturally that the linear mapping canonically associated with A acts from the space $\mathbf{R}^m$ into the space $\mathbf{R}^n$. Also, in accordance with this convention, the expression AX means that the vector X is written in the form of column. The operator $*$ is utilized, as a rule, for conjugation of matrices. So, A^* is the conjugate matrix of A (since we deal only with those matrices whose terms are real numbers, A^* coincides with the transpose of A). If we have a matrix

$$(Z_{kj})_{k \in T_1,\ j \in T_2}$$

whose elements are some random variables, then we put

$$E((Z_{kj})_{k \in T_1,\ j \in T_2}) = (E(Z_{kj}))_{k \in T_1,\ j \in T_2}.$$

The next simple examples give an illustration of the said above.

Example 8. Consider any centered Gaussian random vector

$$X = (X_1, X_2, ..., X_m)^*.$$

Then it is easily seen that

$$cov(X) = E(XX^*).$$

Example 9. Let $X = (X_1, X_2, ..., X_m)^*$ be a standard Gaussian vector and let A be a linear mapping from $\mathbf{R}^m$ into itself, associated with a matrix

$$A = (a_{kj})_{k\in[1,m],\ j\in[1,m]}.$$

Consider the random vector $Y = AX$. Then Y is a centered Gaussian vector and

$$cov(Y) = E(YY^*) = E(AX(AX)^*) = A(E(XX^*))A^* = AA^*.$$

Example 10. Let K be an arbitrary symmetric positive matrix of size $m \times m$. It is well known from the course of linear algebra that K can be represented in the form

$$K = CC^*$$

where C is some matrix of the same size, which is usually called a square root of K and is denoted by $K^{1/2}$. In virtue of Example 9, K is the covariance matrix of the Gaussian random vector $Y = CX$ where X is a standard Gaussian vector in $\mathbf{R}^m$. Note that an analogous fact holds true for infinite (countable) matrices K. For such K, the root C may be chosen to be of triangular form.

Let F be a topological vector space and let H be a subset of F^* separating the points of F. Let X be a random element in the space $(F, B(F, H))$. We recall that the characteristic functional ϕ_X of X is defined by the formula

$$\phi_X(h) = E(exp(ih \circ X)) \qquad (h \in lin(H)).$$

Obviously, we may formulate the definition of a Gaussian random element in the following (equivalent) manner. Namely, we can say that a random element X in $(F, B(F, H))$ is Gaussian if and only if the characteristic functional of X is of the form

$$\phi_X(h) = exp(ia_h - \frac{Q(h,h)}{2}) \qquad (h \in lin(H))$$

where Q denotes the covariance form of X. Since, according to the well-known result from probability theory, the distribution of X is completely determined by ϕ_X (cf. Exercise 5 from Section 6), we see that the covariance of a centered Gaussian element uniquely describes its distribution. So, we can speak of a Gaussian $N(0, Q)$-distributed random element.

So far, we were concerned with H-weak Gaussian random elements. Nevertheless, the most interesting cases appear when random elements in the space $(F, B(F))$ are studied, where $B(F)$ denotes, as usual, the Borel σ-algebra of F (such random elements are called strong ones). In this connection, it is natural to say that a random element X in $(F, B(F))$ is Gaussian if, for any linear functional $h \in F^*$, the corresponding random variable $h \circ X$ turns out to be Gaussian. It should be observed that, for a Polish topological vector space F whose conjugate F^* separates the points of F, the definitions of an F^*-weak Gaussian element and of a strong Gaussian element in F are equivalent. A significant difference between these two notions can be met in the case when a given topological vector space F is nonseparable. Besides, if F^* is poor (for instance, if $F^* = \{0\}$), then the definitions above become useless. In such a situation, it is much more convenient to apply the general notion of a Gaussian random element which is formulated in Exercise 9 to this section.

In virtue of the introduced concepts, we shall say that a probability measure μ on $(F, B(F, H))$ is Gaussian if its characteristic functional ϕ_μ can be represented in the form

$$\phi_\mu(h) = \int_F exp(ih(x))\mu(dx) = exp(ia_h - \frac{Q(h,h)}{2}) \quad (h \in lin(H))$$

where

$$a_h = \int_F h(x)\mu(dx), \quad Q(h,h) = \int_F (h(x) - a_h)^2 \mu(dx).$$

Thus, we see that the notions of a Gaussian element and of a Gaussian measure are closely related. Namely, the distribution of a Gaussian element is a Gaussian measure and, conversely, if μ is a Gaussian measure on F, then the random element X defined on the basic probability space

$$(F, B(F, H), \mu)$$

by the formula

$$X(u) = u \quad (u \in F),$$

turns out to be Gaussian in the space $(F, B(F, H))$. This interaction enables us to transfer directly the results and facts from the theory of Gaussian random elements to Gaussian measures and, conversely, various results and facts from the theory of Gaussian measures can be formulated in terms of Gaussian random elements.

Example 11. Let K be the covariance matrix of a centered Gaussian measure μ in the space $\mathbf{R}^m$ where $m \geq 1$. If $det(K) \neq 0$, then μ is absolutely continuous with respect to the standard Lebesgue measure λ_m on $\mathbf{R}^m$. In this case, it is not hard to find the density p of μ. Namely, we get the following formula:

$$p(x) = \frac{exp((-1/2) < K^{-1}x, x >)}{(2\pi)^{m/2}(det(K))^{1/2}} \qquad (x \in \mathbf{R}^m),$$

where the symbol $<, >$ stands for the canonical inner product in $\mathbf{R}^m$.

We have explained above that a centered Gaussian random element (or, equivalently, a centered Gaussian measure) are completely defined by their covariance. From this point of view, Examples 9 and 10 show that any centered Gaussian element Y in the product space $\mathbf{R}^{\mathbf{N}}$ is equidistributed with some random element AX where X is a standard Gaussian sequence and A is a linear mapping acting from $\mathbf{R}^{\mathbf{N}}$ into itself and canonically associated with a certain infinite triangular matrix. It immediately follows from this fact that every centered Gaussian measure on $(\mathbf{R}^{\mathbf{N}}, B(\mathbf{R}^{\mathbf{N}}))$ belongs to the class $M' = M'(\mathbf{R}^{\mathbf{N}})$ introduced earlier (see Definition 3 from this section). Furthermore, Theorem 2 implies the next auxiliary proposition.

Lemma 6. *Any centered Gaussian random element given in the space $(\mathbf{R}^{\mathbf{N}}, B(\mathbf{R}^{\mathbf{N}}))$ (respectively, any centered Gaussian measure in the same space) is symmetric and unimodal.*

The preceding lemma enables to establish a more general fact. Namely, we have the following statement.

Theorem 3. *Let F be a Polish topological vector space such that F^* separates the points of F, and let Y be an arbitrary centered Gaussian random element in the space $(F, B(F))$. Then Y is a symmetric and unimodal random element. In other words, any centered Gaussian measure on $(F, B(F))$ turns out to be symmetric and unimodal.*

Proof. The cases when

$$F = \mathbf{R}^m \quad (m \geq 1), \qquad F = \mathbf{R}^{\mathbf{N}}$$

were considered earlier. In the case of general F, we can find a countable family $(h_k)_{k \in \mathbf{N}}$ of functionals from F^*, separating the points of F. Now, we define a mapping

$$\Psi : F \rightarrow \mathbf{R}^{\mathbf{N}}$$

by the formula

$$\Psi(x) = (h_k(x))_{k \in \mathbf{N}} \qquad (x \in F).$$

Evidently, Ψ is an injective linear continuous mapping from F into $\mathbf{R}^{\mathbf{N}}$. Under this mapping, the centered Gaussian random element Y is transformed into the centered Gaussian random element $\Psi(Y) = \Psi \circ Y$, and a convex symmetric (with respect to zero) set $D \in B(F)$ is transformed onto the convex symmetric set $\Psi(D) \in B(\mathbf{R}^{\mathbf{N}})$. Then, in view of Lemma 6, for any $t \in [0,1]$ and for each $y \in F$, we obtain the relations

$$P(\{Y \in D + ty\}) = P(\{\Psi(Y) \in \Psi(D + ty)\}) =$$

$$P(\{\Psi(Y) \in \Psi(D) + t\Psi(y)\}) \geq$$

$$P(\{\Psi(Y) \in \Psi(D) + \Psi(y)\}) = P(\{Y \in D + y\})$$

whence it follows that Y is a unimodal random element. Trivially, Y is also symmetric, and Theorem 3 has thus been proved.

We want to conclude this section with a brief discussion of some other properties of Gaussian random elements.

Let F_1 and F_2 be two topological vector spaces, let

$$H_1 \subset F_1^*, \qquad H_2 \subset F_2^*,$$

and suppose that X_1 and X_2 are any two jointly Gaussian random elements in the spaces $(F_1, B(F_1, H_1))$ and $(F_2, B(F_2, H_2))$ respectively. Then, for each $h_1 \in lin(H_1)$ and for each $h_2 \in lin(H_2)$, we have the equality

$$E(exp(i(h_1(X_1) + h_2(X_2)))) =$$

$$exp(ia_{h_1} + ia_{h_2} - \frac{Q_1(h_1,h_1) + 2Q_{1,2}(h_1,h_2) + Q_2(h_2,h_2)}{2})$$

where Q_1 and Q_2 are the covariance forms of X_1 and X_2 respectively, and

$$Q_{1,2} = cov(X_1, X_2)$$

is the mutual covariance of X_1 and X_2 defined by

$$Q_{1,2}(h_1,h_2) = E((h_1(X_1) - a_{h_1})(h_2(X_2) - a_{h_2}))$$

for any $h_1 \in lin(H_1)$ and $h_2 \in lin(H_2)$. Also, it is clear that, for all $h_1 \in lin(H_1)$ and $h_2 \in lin(H_2)$, the equality

$$E(exp(i(h_1(X_1) + h_2(X_2)))) =$$

$$exp(ia_{h_1} - \frac{Q_1(h_1,h_1)}{2}) \times exp(ia_{h_2} - \frac{Q_2(h_2,h_2)}{2})$$

holds true if and only if the random elements X_1 and X_2 are noncorrelated, i.e.

$$Q_{1,2}(h'_1, h'_2) = 0 \qquad (h'_1 \in H_1,\ h'_2 \in H_2).$$

Thus, we may assert that jointly Gaussian random elements X_1 and X_2 are independent if and only if they are noncorrelated (cf. Exercise 7 from Section 6).

We recall that a sequence $(X_n)_{n\in\mathbf{N}}$ of random variables converges in mean square to a random variable X if

$$lim_{n\to\infty} E((X_n - X)^2) = 0.$$

It can easily be verified that if our $(X_n)_{n\in\mathbf{N}}$ is a Gaussian sequence, then its limit X (in the above-mentioned sense) is a Gaussian random variable. Moreover, it turns out that the extended sequence

$$\{X\} \cup \{X_n \ :\ n \in \mathbf{N}\}$$

is jointly Gaussian, too. Thus, we conclude that the closure (with respect to the metric induced by the mean square convergence) of the linear hull of any Gaussian family is also a Gaussian family. This simple observation implies some important consequences. Let us formulate and prove one of them.

Lemma 7. *Let $Y = (Y_t)_{t\in T}$ be a mean square continuous centered Gaussian process given on a separable metric space T (e.g., on $T = [0,1]$). Then there exist a standard Gaussian sequence $(X_k)_{k\in\mathbf{N}}$ and a sequence $(\phi_k)_{k\in\mathbf{N}}$ of real-valued functions on T, such that:*

1) all X_k are measurable with respect to the σ-algebra

$$S' = \sigma((Y_t)_{t\in T}, B(\mathbf{R}));$$

2) all ϕ_k are continuous;

3) for each point $t \in T$, we have

$$Y_t = \sum_{k\in\mathbf{N}} X_k \phi_k(t)$$

where the equality above states that this series converges in mean square and P-almost surely.

Proof. As usual, we denote by $L_2(\Omega)$ the family of all those random variables on (Ω, S) which are square integrable with respect to P. Of course, we identify P-equivalent random variables. Note that $L_2(\Omega)$ is a Hilbert space with respect to the canonical inner product

$$< X, Z > = E(XZ) \qquad (X \in L_2(\Omega),\ Z \in L_2(\Omega)),$$

and the mean square convergence coincides with the convergence in the metric induced by this inner product. Consider the closed vector subspace $G(Y)$ of $L_2(\Omega)$ generated by the given family $(Y_t)_{t \in T}$. Clearly, $G(Y)$ is a Hilbert subspace of $L_2(\Omega)$ and, simultaneously, is a Gaussian family of centered random variables. Moreover, $G(Y)$ is separable since, in view of our assumption, the mapping

$$t \rightarrow Y_t \qquad (t \in T)$$

acting from T into $L_2(\Omega)$ is continuous. Hence we may choose a countable orthonormal basis $(X_k)_{k \in \mathbf{N}}$ in $G(Y)$ which forms a standard Gaussian sequence. Therefore, for any $t \in T$, the random variable Y_t regarded as an element of $G(Y)$ can be expressed as

$$Y_t = \sum_{k \in \mathbf{N}} X_k E(Y_t X_k)$$

where the above series converges in the metric of $L_2(\Omega)$, i.e. converges in mean square. But this series converges also almost surely because $(X_k)_{k \in \mathbf{N}}$ is a standard Gaussian sequence and

$$\sum_{k \in \mathbf{N}} (E(Y_t X_k))^2 < +\infty$$

for all $t \in T$ (cf. Example 1 from Section 7). Further, for each $k \in \mathbf{N}$, let us define a function

$$\phi_k \ : \ T \rightarrow \mathbf{R}$$

by putting

$$\phi_k(t) = E(Y_t X_k) \qquad (t \in T).$$

Then, for all $t \in T$ and $s \in T$ and for each natural number k, the inequality

$$(\phi_k(t) - \phi_k(s))^2 \leq E((Y_t - Y_s)^2)$$

holds true. But, according to the assumption of the lemma, our process Y is mean square continuous whence it follows that all functions ϕ_k are continuous. This completes the proof.

In general, any representation of a given process in the form of the series of mutually independent summands is useful in various problems of probability theory. This circumstance will be illustrated in the next sections of our book. Of course, in order to have an opportunity to apply such representations, we need some regularity properties of the original process (e.g., as in the preceding lemma).

In connection with various regularity properties of stochastic processes, it is reasonable to recall here the concept of a separable process.

Let Y be some stochastic process defined on a metric (or, more generally, topological) space T. We say that this process is separable if there exist a countable set $Q \subset T$ and a set $\Omega_0 \subset \Omega$ with $P(\Omega_0) = 1$, such that, for each $t \in T$ and for any $\omega \in \Omega_0$, a sequence $(q_n)_{n \in \mathbf{N}} \subset Q$ can be found satisfying the relations

$$lim_{n \to \infty} q_n = t, \quad lim_{n \to \infty} Y(\omega, q_n) = Y(\omega, t).$$

In particular, the set Q turns out to be dense in T. This set is usually called a set of the separability of Y.

Note that the separability condition does not essentially restrict the class of stochastic processes. Indeed, according to the fundamental theorem of the general theory of stochastic processes (see, e.g., [57], [155] or [171]), any process defined on a separable metric space admits a separable modification. This deep result enables us to avoid many difficulties connected with the question of measurability of various sets and functions under consideration. Also, this result can be deduced from another deep statement concerning the existence of multiplicative liftings (cf. [171]).

For further information about the topic of this section, see [20], [22], [23], [24], [25], [65], [66], [67], [134], [135], [137], [200].

EXERCISES

1. Let F be a Polish topological vector space having the property that, for each compact set $K \subset F$, the closed convex hull of K is compact, too. Demonstrate that, for a probability measure μ on $(F, B(F))$, the following two assertions are equivalent:

a) μ is unimodal;

b) $\mu(D + ty) \geq \mu(D + y)$ whenever $t \in [0, 1]$, $y \in F$ and D is a compact convex symmetric (with respect to zero) subset of F.

2. Let F be a separable Frechet space and let μ be a probability measure on $(F, B(F))$. Show that μ is unimodal if and only if the Anderson

inequality holds true for all convex symmetric (with respect to zero) bodies D in F. Moreover, in the case $F = \mathbf{R}^m$ where $m \geq 1$, show that it suffices to take D only from the family of all compact, strongly convex and symmetric bodies in F (apply the result of Exercise 10 from Section 2).

3. Prove that any unimodal (in the sense of Anderson) probability measure on $(\mathbf{R}, \mathcal{B}(\mathbf{R}))$ is symmetric. Generalize this fact to the case of a Polish topological vector space F such that F^* separates the points of F.

4. Demonstrate that a probability measure μ on $(\mathbf{R}, \mathcal{B}(\mathbf{R}))$ is unimodal in the sense of Khinchin (with a mode at a point $a \in \mathbf{R}$) if and only if the distribution function of μ is lower convex on the interval $]-\infty, a[$ and upper convex on the interval $]a, +\infty[$.

5. Give detailed proofs of Lemmas 1 and 4, and of Theorem 1.

6. Find a suitable analogue of Lemma 2 in the case of $\mathbf{R}^m$, where $m > 1$, and prove it.

7. Deduce from Theorem 1 the following statement: the convolution of any finite family of symmetric unimodal probability measures on $(\mathbf{R}, \mathcal{B}(\mathbf{R}))$ is a symmetric unimodal probability measure (this result is due to Wintner).

8. Let F be a Polish topological vector space, let H be a subset of F^* separating the points of F and let X be a centered Gaussian random element in the space $(F, \mathcal{B}(F, H))$. Demonstrate that X is simultaneously a random element in the space $(F, \mathcal{B}(F))$ and, for each functional $h \in F^*$, the random variable $h \circ X$ is centered Gaussian.

9. Let (F, Ξ) be a measurable vector space, i.e. F is a vector space and Ξ is a σ-algebra of subsets of F, compatible with the vector structure of F (this means that the vector operations in F are measurable). Suppose also that X is a random element in the space (F, Ξ). We shall say that X is centered Gaussian in (F, Ξ) if, for any two random elements X_1 and X_2 in the same space, which are independent copies of X, the random elements

$$Y_1 = (1/2)^{1/2}(X_1 + X_2), \quad Y_2 = (1/2)^{1/2}(X_1 - X_2)$$

are also independent copies of X.

Show that, for a space F satisfying the conditions of Exercise 8, the definition above is equivalent to the standard definition of a centered Gaussian random element in $(F, \mathcal{B}(F))$.

In addition to this, show that if X is a centered Gaussian random element in the space (F, Ξ), then, for each Ξ-measurable linear functional h, the composition $h \circ X$ turns out to be a centered Gaussian random variable.

10. Give a detailed proof of Theorem 3.

11. Demonstrate (e.g., for $T = [0, 1]$) that Lemma 7 can be strengthened under the additional assumption in its formulation, saying that P-almost all trajectories of a given process $Y = (Y_t)_{t \in T}$ are continuous. In this case, we have

$$P(\{lim_{n\to\infty} sup_{t\in T} |Y_t - \sum_{0\leq k\leq n} X_k\phi_k(t)| = 0\}) = 1.$$

12. Let (Ω, S, P) be a probability space, let

$$X \ : \Omega \to \mathbf{R}$$

be a random variable on this space and let

$$\psi \ : \mathbf{R} \to \mathbf{R}$$

be a Borel mapping. Prove that

$$\int_{\Omega} \psi(X(\omega))P(d\omega) = \int_{\mathbf{R}} \psi(x)\mu_X(dx)$$

whenever the above-mentioned integrals exist.

This simple formula is rather useful in many situations and possesses various applications in probability theory. In particular, for each natural number n, let us take as ψ the function

$$\psi_n \ : \mathbf{R} \to \mathbf{R}$$

defined by the formula

$$\psi_n(x) = x^n \qquad (x \in \mathbf{R}).$$

Then the integral

$$\int_{\Omega} \psi_n(X(\omega))P(d\omega) = \int_{\mathbf{R}} x^n \mu_X(dx)$$

is called the moment (of order n) of the random variable X (of course, under the assumption that this integral exists).

Obviously, the moment of order 0 exists for any X and is equal to 1.

Further, we have:

a) the moment of order 1 coincides with the mathematical expectation of X;

b) if X is centered, then the moment of order 2 coincides with the variance of X.

Note also that the family of all moments of a given random variable X yields an essential information about the behaviour of X. In this connection, let us recall the formulation of the classical problem of moments: is it possible to reconstruct the distribution of X by means of all its moments? Under some additional assumptions, this problem admits a positive solution, i.e. the family of all moments of X determines μ_X uniquely (see, e.g., [74] or [180]).

13. Denote by the symbol J_0 the family of all those probability distributions on $\mathbf{R}$ which are locally absolutely continuous with respect to the standard Lebesgue measure λ_1 on $\mathbf{R} \setminus \{0\}$. Also, denote by the symbol J_1 the family of all those unimodal probability distributions on $\mathbf{R}$ whose modes coincide with zero.

Let Φ be an arbitrary distribution function on $\mathbf{R}$ corresponding to some measure from J_0. We define a function

$$F : \mathbf{R} \rightarrow \mathbf{R}$$

by the formula:

$$F(x) = -\int_{-\infty}^{x} du \int_{-\infty}^{u} \frac{\Phi'(v)}{v} dv \qquad (x < 0),$$

$$F(x) = 1 - \int_{x}^{+\infty} du \int_{u}^{+\infty} \frac{\Phi'(v)}{v} dv \qquad (x > 0).$$

Show that F is an unimodal distribution function (with a mode at zero). Show also that

$$F(x) - xF'(x) = \Phi(x)$$

for λ_1-almost all points $x \in \mathbf{R}$.

Conclude that there exists a canonical bijection between the family of all distribution functions on $\mathbf{R}$ corresponding to probability measures from the class J_0 and the family of all distribution functions on $\mathbf{R}$ corresponding to probability measures from the class J_1 (this result is due to Khinchin).

11. Oscillation phenomena and extensions of measures

It is well known that random motion of various objects can be regarded as a mathematical model of certain physical processes and is investigated by the techniques of modern probability theory. The trajectories of such processes may be rather complicated. A characteristic feature of these trajectories is their significant oscillation. In particular, this phenomenon can easily be discovered by watching realizations of images, under linear transformations, of sequences of independent random variables (assuming that those images are bounded but not converging to zero). The classical example of such a situation is provided by the widely known law of iterated logarithm (see, e.g., [68], [73], [118] or [187]). Note also that strange oscillation properties are reflected in some extraordinary effects for the images of countable products of probability measures, under their linear transformations.

The above-mentioned phenomena form a base of the material presented below and, which is much more important, find their natural interpretation in the general problem of extending a probability measure on the cylindrical σ-algebra of the space l_∞ to a probability measure on the Borel σ-algebra of this space. The unimodality of a given measure (and, hence, the Anderson inequality) turns out to be a link here. So, we will be able to demonstrate in this section close relationships between the measurability problems and the convexity inequalities.

Let $X = (X_n)_{n \in \mathbf{N}}$ be a sequence of independent random variables and let

$$A = (a_{nk})_{n \in \mathbf{N},\ k \in \mathbf{N}}$$

be an infinite triangular matrix whose columns belong to the classical Banach space c_0. More precisely, suppose that:

1) $a_{nk} \in \mathbf{R}$ for all natural numbers n and k;

2) $a_{nk} = 0$ if $k > n$;

3) for each $k \in \mathbf{N}$, we have

$$lim_{n \to \infty} a_{nk} = 0.$$

The class of all matrices A satisfying conditions 1) - 3) will be denoted by the symbol $V_0 = V_0(\mathbf{R}^{\mathbf{N}})$. Obviously, any matrix $A \in V_0$ induces the corresponding linear mapping (operator) A from $\mathbf{R}^{\mathbf{N}}$ into $\mathbf{R}^{\mathbf{N}}$, and we get

$$Y = A \circ X = A(X) = AX$$

where $Y = (Y_n)_{n \in \mathbf{N}}$ is a random sequence such that

$$Y_n = \sum_{0 \leq k \leq n} a_{nk} X_k \qquad (n \in \mathbf{N}).$$

Note that an analogous transformation is sensible for a much wider class of matrices (not necessarily fulfilling condition 2)). Indeed, it suffices to assume that, for each $n \in \mathbf{N}$, the series

$$\sum_{k \in \mathbf{N}} a_{nk} X_k$$

is convergent P-almost surely. However, we are not interested in such a generalization and deal only with the class of infinite triangular matrices mentioned above.

The family of all random sequences Y obtained in this manner will be denoted by the symbol $W_0 = W_0(\mathbf{N})$. Let us emphasize once more that all original random sequences X are assumed to be defined on a fixed probability space (Ω, S, P). This assumption does not restrict essentially the generality of our considerations below.

Definition 1. Let $Z = (Z_n)_{n \in \mathbf{N}}$ be a sequence of random variables. A real number $\alpha \geq 0$ (possibly, equal to $+\infty$) is called an oscillation constant for Z if the relation

$$P(\{limsup_{n \to \infty} |Z_n| = \alpha\}) = 1$$

holds true.

In this case, we shall also write $\alpha = \alpha(Z)$, stressing especially the circumstance that α depends on Z. Let us give some simple examples illustrating this notion.

Example 1. Let $a \in \mathbf{R}$ and let $Z_n = a$ for all natural numbers n. Evidently, for $Z = (Z_n)_{n \in \mathbf{N}}$, we have $\alpha(Z) = a$.

Example 2. Let $Y = (Y_n)_{n \in \mathbf{N}}$ be a sequence of independent random variables, such that

$$ran(Y_n) = \{-1, 1\} \qquad (n \in \mathbf{N})$$

and, in addition to this,

$$P(\{Y_n = 1\}) = P(\{Y_n = -1\}) = 1/2 \qquad (n \in \mathbf{N}).$$

Then it can easily be seen that $\alpha(Y) = 1$.

Example 3. Let $(X_n)_{n\geq 1}$ be a sequence of independent and equidistributed random variables, such that

$$E(X_1) = 0, \quad E(X_1^2) = 1.$$

Define a new random sequence $Y = (Y_n)_{n\geq 1}$ by putting

$$Y_n = \frac{X_1 + X_2 + \ldots + X_n}{a_n} \qquad (n \geq 1),$$

where $(a_n)_{n\geq 1}$ is some fixed sequence of real numbers.

If $a_n = n$ for all $n \geq 1$, then, according to the strengthened law of large numbers (due to Kolmogorov), we have $\alpha(Y) = 0$.

If $a_n = (n \cdot ln(ln(n)))^{1/2}$ for all $n \geq 3$, then, according to the law of iterated logarithm (due to Hartman and Wintner), we have $\alpha(Y) = 2^{1/2}$.

If $a_n = 1$ for all $n \geq 1$, then $\alpha(Y) = +\infty$.

In connection with these facts, see e.g. [162].

Example 4. Let $Y = (Y_n)_{n\in\mathbf{N}}$ be a sequence of independent Gaussian random variables, such that each Y_n is $N(0, \sigma_n^2)$-distributed.

If, for any $\varepsilon > 0$, the series

$$\sum_{n\in\mathbf{N}} exp(-\varepsilon/\sigma_n^2)$$

is convergent, then $\alpha(Y) = 0$.

If there exists an $\varepsilon > 0$ for which the above-mentioned series is convergent and there exists an $\varepsilon' > 0$ for which the corresponding series is divergent, then $\alpha(Y) \in]0, +\infty[$.

In connection with Example 4, see e.g. [196], [197] and Exercise 1 to this section.

Example 5. Let $(b_k)_{k\in\mathbf{N}}$ be a sequence of real numbers, such that

$$\{b_k \; : \; k \in \mathbf{N}\} \subset]0, +\infty[, \quad \sum_{k\in\mathbf{N}} b_k^2 < +\infty.$$

Define an infinite matrix $A = (a_{nk})_{n\in\mathbf{N},\ k\in\mathbf{N}}$ by putting

$$a_{nk} = b_k \quad (k \leq n), \qquad a_{nk} = 0 \quad (k > n).$$

Further, consider the random sequence $Y = AX$ where $X = (X_n)_{n\in\mathbf{N}}$ is a standard Gaussian sequence, i.e. a sequence of independent random variables each of which is $N(0,1)$-distributed. Then, obviously,

$$Y_n = \sum_{0\leq k\leq n} b_k X_k \qquad (n \in \mathbf{N})$$

and the random sequence $Y = (Y_n)_{n\in\mathbf{N}}$ is convergent P-almost surely to some $N(0, \sum_{k\in\mathbf{N}} b_k^2)$-distributed random variable. Now, it is clear that, for each number $\beta \in [0, +\infty]$, the relation

$$P(\{limsup_{n\to\infty}|Y_n| = \beta\}) = 0$$

holds true whence it follows that Y possesses no oscillation constant.

Even if we know that there exists an oscillation constant for a given random sequence, it is hard, in general, to find its precise value (excluding some trivial cases similar to those described in Examples 1 and 2 above). Moreover, less difficult problems concerning the validity of relations

$$\alpha = 0, \qquad \alpha \in]0, +\infty[, \qquad \alpha = +\infty,$$

for an oscillation constant $\alpha = \alpha(X)$, also are far from trivial ones. However, the fact of the existence of an oscillation constant turns out to be rather useful in many concrete situations.

In order to demonstrate this, we need the classical zero-one law of Kolmogorov (cf., e.g., [121], [155], [180] or [197]). Let (Ω, S, P) be a basic probability space and let $(X_n)_{n\in\mathbf{N}}$ be a sequence of independent random variables defined on this space. For each natural number m, we denote by $S_{(m)}$ the σ-algebra of subsets of Ω, generated by the partial sequence $(X_n)_{n>m}$. Evidently, $S_{(m)} \subset S$. We also put

$$S_{(\infty)} = \cap_{m\in\mathbf{N}} S_{(m)}.$$

Since the intersection of a family of σ-algebras is a σ-algebra, too, we see that $S_{(\infty)}$ turns out to be a σ-algebra in Ω. It will be called the remainder (or the residual) σ-algebra for $(X_n)_{n\in\mathbf{N}}$. This terminology can be explained by the fact that, for each $m \in \mathbf{N}$, any event belonging to $S_{(\infty)}$ is independent with respect to the random variables X_0, X_1, ..., X_m. The standard formulation of the zero-one law is as follows:

$$(\forall B \in S_{(\infty)})(P(B) = 0 \ or \ P(B) = 1).$$

Let us present a brief proof of this fundamental fact. Suppose that $B \in S_{(\infty)}$ and $P(B) > 0$. We have to show that $P(B) = 1$. For each $m \in \mathbf{N}$, let us put:

$S^{(m)}$ = the σ-algebra generated by $(X_0, X_1, ..., X_m)$;

$S^{(\infty)}$ = the σ-algebra generated by $\cup_{m\in\mathbf{N}} S^{(m)}$.

Note that

$$m \leq m' \Rightarrow S^{(m)} \subset S^{(m')}.$$

Hence the class $\cup_{m\in\mathbf{N}} S^{(m)}$ is an algebra of subsets of Ω, contained in S. Now, define on $S^{(\infty)}$ a new probability measure Q by the formula

$$Q(C) = P(C \cap B)/P(B) \qquad (C \in S^{(\infty)}).$$

Taking into account the fact that, for any $m \in \mathbf{N}$, the σ-algebras $S_{(\infty)}$ and $S^{(m)}$ are independent, we obtain

$$P(C) = Q(C) \qquad (C \in \cup_{m\in\mathbf{N}} S^{(m)}).$$

Therefore, in virtue of the classical Carathéodory theorem on the existence and uniqueness of an extension of a measure, we conclude that

$$P(C) = Q(C) \qquad (C \in S^{(\infty)}).$$

But we have the obvious inclusion

$$S_{(\infty)} \subset S^{(\infty)}$$

whence it follows that

$$P(B) = Q(B) = P(B \cap B)/P(B) = 1.$$

This completes the proof of the zero-one law.

The following auxiliary statement may be regarded as another (equivalent) formulation of the zero-one law.

Lemma 1. *Let Z be a positive extended random variable measurable with respect to the remainder σ-algebra $S_{(\infty)}$ introduced above. Then Z is degenerate; in other words, there exists a constant $\beta \in [0, +\infty]$ for which we have*

$$P(\{Z = \beta\}) = 1.$$

The next auxiliary statement justifies our especial attention to the class $W_0 = W_0(\mathbf{N})$.

Lemma 2. *Let a sequence Y of random variables belong to the class W_0. Then there exists an oscillation constant for Y.*

Proof. Let $Y \in W_0$, i.e. there is a matrix $A \in V_0$ and a sequence $X = (X_n)_{n \in \mathbf{N}}$ of independent random variables, such that $Y = AX$. Condition 3), for matrices from V_0, implies that the relation

$$lim_{n \to \infty} \sum_{0 \leq k \leq m} a_{nk} X_k(\omega) = 0$$

holds true whenever $m \in \mathbf{N}$ and $\omega \in \Omega$. Consequently, for all $m \in \mathbf{N}$ and for all $\omega \in \Omega$, we get

$$limsup_{n \to \infty} |Y_n(\omega)| = limsup_{n \to \infty} |\sum_{0 \leq k \leq m} a_{nk} X_k(\omega) + \sum_{k > m} a_{nk} X_k(\omega)| =$$

$$limsup_{n \to \infty} |\sum_{k > m} a_{nk} X_k(\omega)|.$$

Now, it is clear that the random variable

$$Z = limsup_{n \to \infty} |Y_n|$$

is measurable with respect to the remainder σ-algebra $S_{(\infty)}$ associated with the random sequence $(X_n)_{n \in \mathbf{N}}$. Since

$$limsup_{n \to \infty} |Y_n| \geq 0,$$

we obtain, in view of Lemma 1, the existence of an oscillation constant for our random sequence Y.

There are numerous functionals defined on W_0 which are degenerate and, hence, determine some real numbers analogous to oscillation constants. The description and investigation of such functionals is of interest but is beyond the scope of our book. Here we are going to consider only two functionals of this kind:

$$\alpha_-(Y) = liminf_{n \to \infty} Y_n, \quad \alpha_+(Y) = limsup_{n \to \infty} Y_n,$$

where $Y \in W_0$. In general, there are no nontrivial relations between α_- and α_+, but it is evident that if α is an oscillation constant, then

$$\alpha = max(\alpha_+, |\alpha_-|).$$

Suppose now that Y is symmetric (e.g., this is valid when a random sequence X for which $Y = AX$, is a sequence of independent symmetric random variables). Then we have the equalities

$$\alpha_+ = -\alpha_- = \alpha.$$

Further, Example 5 considered earlier shows us that if a random sequence Y does not belong to the class W_0, then it can happen that Y possesses no oscillation constant. It must also be mentioned that the following equivalence relations are true:

$$\alpha(Y) = 0 \Leftrightarrow P(\{Y \in c_0\}) = 1,$$

$$\alpha(Y) = +\infty \Leftrightarrow P(\{Y \notin l_\infty\}) = 1,$$

$$\alpha(Y) \in]0, +\infty[\Leftrightarrow P(\{Y \in l_\infty \setminus c_0\}) = 1.$$

For our purposes, the third case is of primary interest. If this case holds and our random sequence Y is symmetric, then almost all realizations of Y are infinitely many times in arbitrarily small neighbourhoods of the points $-\alpha$ and α. The same can be expressed in the following manner: $-\alpha$ and α are limit points of almost all realizations of Y, and no real $\beta \notin [-\alpha, \alpha]$ is a limit point of realizations of Y with strictly positive probability. Such a behaviour of trajectories is completely natural for a random sequence indicated in Example 2 and is implied by the discrete structure of this sequence. But, for other (nondiscrete) types of random sequences, the oscillation phenomenon is much more interesting and complicated. First of all, this remark can be addressed to one important subclass of W_0 consisting of Gaussian random sequences. For such random sequences, the oscillation properties of realizations admit a detailed analysis, and we shall thoroughly discuss these properties in the sequel.

We wish to begin with some conditions which guarantee that a given centered Gaussian random sequence belongs to the class W_0.

Let $Y = (Y_n)_{n \in \mathbf{N}}$ be an $N(0, K)$-distributed Gaussian random sequence where K denotes the covariance matrix of Y. Then there always exist a standard Gaussian sequence X and an infinite triangular matrix A, such that $Y = AX$. Hence $Y \in W_0$ if and only if $A \in V_0$. Further, since $K = AA^*$ where A^* denotes the transpose of A, the problem can be reduced to the study of columns of $A = K^{1/2}$. Investigation in this direction leads, of course, to a criterion of belonging Y to W_0, but such a criterion is not suitable for us. An essentially different approach based on the asimptotic behaviour of elements of the main diagonal of K seems to be much more convenient and helpful. To see this, let us denote

$$\sigma_n^2 = E(Y_n^2) \quad (n \in \mathbf{N}).$$

We have the next auxiliary proposition which will be useful below.

Lemma 3. *Let $Y = (Y_n)_{n \in \mathbf{N}}$ be a centered Gaussian random sequence. If the equality*

$$lim_{n \to \infty} \sigma_n = 0$$

is valid, then Y belongs to W_0.

Proof. Indeed, let an infinite triangular matrix A and a standard Gaussian sequence X be such that $Y = AX$. Then, for any $n \in \mathbf{N}$ and for each natural number $k \leq n$, we have

$$a_{nk}^2 \leq \sum_{0 \leq j \leq n} a_{nj}^2 = \sigma_n^2$$

whence it follows that

$$lim_{n \to \infty} a_{nk} = 0,$$

which completes the proof.

In the case of Gaussian random sequences, Lemma 2 can be essentially strengthened. In order to present the strong version of this lemma, we need some notation. Let p be a natural number, let $\mathbf{R}^p$ denote, as usual, the p-dimensional Euclidean space and let $(x_n)_{n \in \mathbf{N}}$ be a sequence of points from this space. Then the set of all limit points of the above-mentioned sequence will be denoted by the symbol $D((x_n)_{n \in \mathbf{N}})$. With this notation, we can formulate and prove the following result (see [45]).

Theorem 1. *Let $Y = (Y_n)_{n \in \mathbf{N}}$ be a centered Gaussian random sequence and suppose that*

$$lim_{n \to \infty} E(Y_n^2) = 0,$$

$$P(\{sup_{n \in \mathbf{N}} |Y_n| < +\infty\}) = 1.$$

Then, for P-almost all $\omega \in \Omega$, the relation

$$D((Y_n(\omega))_{n \in \mathbf{N}}) = [-\alpha, \alpha]$$

holds true, where $\alpha = \alpha(Y)$ is the oscillation constant for Y. In other words, the set of all limit points of almost all realizations of Y is fixed and coincides with the closed bounded interval $[-\alpha, \alpha]$.

Proof. The argument below utilizes one geometric trick and appeals to the corresponding properties of the Euclidean space $\mathbf{R}^p$ where $p > 1$ (cf. [77]). As usual, we denote by the symbol $B_{p,r}(0)$ the p-dimensional closed ball in this space, whose radius equals r and whose centre coincides with 0, i.e.

$$B_{p,r}(0) = \{x \in \mathbf{R}^p \ : \ ||x|| \leq r\}.$$

Accordingly, the symbol $S_{p-1,r}(0)$ will denote the boundary of this ball, i.e the $(p-1)$-dimensional sphere with the same centre and radius. The standard inner product in $\mathbf{R}^p$ will be denoted by $<,>$.

Consider now p independent copies

$$Y^{(i)} = (Y_n^{(i)})_{n \in \mathbf{N}} \qquad (1 \leq i \leq p)$$

of our random sequence Y. Obviously, these copies can be defined on the basic probability space (Ω, S, P) and, in addition, we may put

$$Y^{(1)} = Y = (Y_n)_{n \in \mathbf{N}}.$$

Starting with the copies $Y^{(1)}$, $Y^{(2)}$, ..., $Y^{(p)}$, we introduce a new sequence $Y' = (Y'_n)_{n \in \mathbf{N}}$ of p-dimensional random vectors, by putting

$$Y'_n = (Y_n^{(1)}, \ Y_n^{(2)}, \ ..., \ Y_n^{(p)}).$$

Clearly, Y' is a centered Gaussian sequence in $\mathbf{R}^p$.

Applying our assumption $lim_{n \to \infty} E(Y_n^2) = 0$ and taking into account Lemma 3, we see that $Y \in W_0$. Hence, in view of Lemma 2, there exists an oscillation constant

$$\alpha(Y) = \alpha \in [0, +\infty].$$

Since $sup_{n \in \mathbf{N}} |Y_n| < +\infty$ almost surely, we conclude that α is finite. Also, assuming that $\alpha = 0$, we easily infer that $Y \in c_0$ with P-probability one, whence it follows that, for almost all ω from Ω, the equality

$$D((Y_n(\omega))_{n \in \mathbf{N}}) = \{0\}$$

is valid. Excluding this trivial case, we may suppose in our further considerations that

$$\alpha \in]0, +\infty[.$$

Let us demonstrate that, for $p \geq 1$, the following relation holds true:

$$(1) \quad P(\{\omega \in \Omega \ : \ D((Y'_n(\omega))_{n \in \mathbf{N}}) = B_{p,\alpha}(0)\}) = 1.$$

Clearly, by applying the corresponding homothety, we may reduce our considerations to the case $\alpha = 1$. In other words, we have to show that if $\alpha = 1$, then the set of all limit points of almost all realizations of Y' coincides with the closed init ball $B_{p,1}(0)$. In particular, for $p = 1$, we get the required result. The approach presented here utilizes good geometric properties of the unit sphere $S_{p-1,1}(0)$ where $p > 1$.

Fix a vector

$$h = (h_1, h_2, ..., h_p) \in S_{p-1,1}(0)$$

and consider the sequence

$$Z_h = (< h, Y'_n >)_{n \in \mathbf{N}}$$

of random variables. Note that Z_h is a centered Gaussian sequence whose full characterization can be obtained by the covariance matrix

$$cov(Z_h) = (E(< h, Y'_n > \cdot < h, Y'_m >))_{n \in \mathbf{N},\ m \in \mathbf{N}}.$$

For any natural numbers n and m, we have

$$E(< h, Y'_n > \cdot < h, Y'_m >) = \sum_{1 \leq i \leq p,\ 1 \leq j \leq p} h_i h_j E(Y_n^{(i)} Y_m^{(j)}).$$

But if $i \neq j$, then the random variables $Y_n^{(i)}$ and $Y_m^{(j)}$ are independent, and

$$E(Y_n^{(i)} Y_m^{(i)}) = E(Y_n Y_m)$$

for all natural numbers $i \in [1, p]$. Hence

$$E(< h, Y'_n > \cdot < h, Y'_m >) = \sum_{1 \leq i \leq p} h_i^2 E(Y_n^{(i)} Y_m^{(i)}) = E(Y_n Y_m).$$

Therefore, for any $h \in S_{p-1,1}(0)$, the relation

$$cov(Z_h) = cov(Y)$$

is fulfilled. This means that Z_h is a copy of Y and, consequently,

$$\alpha(Z_h) = \alpha(Y) = 1.$$

Now, fix a $\delta > 0$. Since $S_{p-1,1}(0)$ is compact, there exist a natural number $m = m(\delta)$ and a family

$$\{g_1, g_2, ..., g_m\} \subset S_{p-1,1}(0)$$

such that, for any vector $z \in \mathbf{R}^p$, we have

$$max_{1 \leq i \leq m} | < g_i, z > | \geq (1 - \delta)||z||.$$

It easily follows from this fact that the relations

$$(1 - \delta) limsup_{n \to \infty} ||Y'_n|| \leq limsup_{n \to \infty} max_{1 \leq i \leq m} | < g_i, Y'_n > | =$$

$$max_{1 \leq i \leq m} limsup_{n \to \infty} | < g_i, Y'_n > | = max_{1 \leq i \leq m} \alpha(Z_{g_i}) = 1$$

are satisfied with P-probability one. Tending δ to zero, we get

$$(2) \quad P(\{limsup_{n \to \infty} ||Y'_n|| \leq 1\}) = 1,$$

which immediately yields the inclusion

$$(3) \quad D((Y'_n(\omega))_{n \in \mathbf{N}}) \subset B_{p,1}(0)$$

for P-almost all $\omega \in \Omega$. Now, let us show that the inclusion

$$(4) \quad S_{p-1,1}(0) \subset D((Y'_n(\omega))_{n \in \mathbf{N}})$$

holds true, too, for P-almost all $\omega \in \Omega$. In order to demonstrate this, recall that the random sequence Z_h is symmetric, which implies

$$\alpha_+(Z_h) = \alpha(Z_h) = 1.$$

Therefore,

$$P(\{limsup_{n \to \infty} < h, Y'_n > = 1\}) = 1.$$

We thus see that, for any $h \in S_{p-1,1}(0)$, there exists a set $\Omega_h \subset \Omega$ with $P(\Omega_h) = 1$ having the property that, for each $\varepsilon > 0$, the following inclusions are valid:

$$\Omega_h \subset \{\omega \ : < h, Y'_n(\omega) > \geq 1 - \varepsilon \ for \ infinitely \ many \ n\};$$

$$\Omega_h \subset \{\omega : ||Y'_n(\omega)||^2 \leq 1 + \varepsilon \ for \ sufficiently \ large \ n\}.$$

Hence, if $\omega \in \Omega_h$ and $\varepsilon > 0$, the relation

$$||Y'_n(\omega) - h||^2 = ||Y'_n(\omega)||^2 - 2 < h, Y'_n(\omega) > + ||h||^2 \leq 3\varepsilon$$

is fulfilled for infinitely many n. But this means, for $\omega \in \Omega_h$, that

$$h \in D((Y'_n(\omega))_{n \in \mathbf{N}}).$$

Now, choose a countable subset Q of the sphere $S_{p-1,1}(0)$, everywhere dense in it. Then, in virtue of the said above, we may write

$$Q \subset D((Y'_n(\omega))_{n\in\mathbf{N}}) \qquad (\omega \in \cap_{h\in Q}\Omega_h).$$

Taking into account the equality

$$P(\cap_{h\in Q}\Omega_h) = 1,$$

we conclude that relation (4) is true.

Finally, consider the canonical projection

$$Pr : \mathbf{R}^p \to \mathbf{R}^{p-1}$$

defined by

$$Pr((x_1, x_2, ..., x_p)) = (x_1, x_2, ..., x_{p-1})$$

for any $x = (x_1, x_2, ..., x_p) \in \mathbf{R}^p$ (we assume here that $p \geq 2$). So we have

$$Pr(S_{p-1,1}(0)) = B_{p-1,1}(0),$$

$$Pr((Y_n^{(1)}, Y_n^{(2)}, ..., Y_n^{(p)})) = (Y_n^{(1)}, Y_n^{(2)}, ..., Y_n^{(p-1)}),$$

and, according to relations (4) and (3), we obtain

$$B_{p-1,1}(0) \subset D((Y_n^{(1)}(\omega), ..., Y_n^{(p-1)}(\omega))_{n\in\mathbf{N}}),$$

$$D((Y_n^{(1)}(\omega), ..., Y_n^{(p-1)}(\omega))_{n\in\mathbf{N}}) \subset B_{p-1,1}(0)$$

for P-almost all $\omega \in \Omega$. This implies the required relation (1) and the theorem has thus been proved.

The assumption

$$lim_{n\to\infty}E(Y_n^2) = 0$$

can be replaced by a weaker assumption, namely, by $Y \in W_0$. But, as shown in Exercise 3, it yields nothing new, because (in our situation) these two assumptions turn out to be equivalent.

Also, the following natural question arises: is it essential in the formulation of Theorem 1 that our random sequence Y is Gaussian? We are not going to discuss all aspects of this question and refer the reader to the extensive survey [108]. Nevertheless, it is reasonable to mention here that rather simple arguments based on the classical Fubini theorem enable us to extend Theorem 1 to a much wider class of random sequences Y.

Example 6. Let us consider a random sequence

$$X = (X_k)_{k \in \mathbf{N}} = (Z_k Z'_k)_{k \in \mathbf{N}}$$

where $Z = (Z_k)_{k \in \mathbf{N}}$ is a standard Gaussian sequence and $Z' = (Z'_k)_{k \in \mathbf{N}}$ is a sequence of independent random variables, such that Z and Z' are independent, too. If $A \in V_0$ and $Y = AX$, then the assertion of Theorem 1 holds true for Y. For example, we may take as $(X_k)_{k \in \mathbf{N}}$ any sequence of independent symmetric random variables with stable distributions.

Thus, we have established that if a bounded (with P-probability one) centered Gaussian sequence Y from the class W_0 is not convergent to zero, then its typical realizations oscillate densely in some nondegenerate closed bounded interval, i.e. in $[-\alpha(Y), \alpha(Y)]$, and have no limit points outside this interval. Such a behaviour of realizations shows us the complicated structure of the set of typical realizations of Y and, in addition, indicates that this set is essentially infinite-dimensional. In our considerations below, we give a more precise description of this phenomenon.

Let us denote by M_0 the class of all those probability measures on $(\mathbf{R}^{\mathbf{N}}, B(\mathbf{R}^{\mathbf{N}}))$ which are distributions of random sequences from the class W_0. Note that M_0 can be described in the following manner (cf. Section 6). Take any sequence $(\nu_n)_{n \in \mathbf{N}}$ of probability measures on $(\mathbf{R}, B(\mathbf{R}))$ and put

$$\nu = \prod_{n \in \mathbf{N}} \nu_n .$$

Further, for an arbitrary Borel mapping

$$\psi : \mathbf{R}^{\mathbf{N}} \to \mathbf{R}^{\mathbf{N}},$$

define

$$\nu_\psi = \nu \circ \psi^{-1}.$$

Then a measure μ belongs to the class M_0 if and only if

$$\mu = \nu_A = \nu \circ A^{-1}$$

for some linear mapping $A : \mathbf{R}^{\mathbf{N}} \to \mathbf{R}^{\mathbf{N}}$ whose matrix belongs to V_0. Here we wish to stress once more the interaction between the class W_0 of random sequences and the class M_0 of probability measures on $(\mathbf{R}^{\mathbf{N}}, B(\mathbf{R}^{\mathbf{N}}))$. In particular, from Lemma 2 we obtain at once the next auxiliary proposition.

Lemma 4. *Let μ be a measure from M_0. Then there exists a number $\alpha = \alpha(\mu) \in [0, +\infty]$ for which*

$$\mu(\{x \in \mathbf{R}^{\mathbf{N}} \; : \; limsup_{n \to \infty}|x_n| = \alpha\}) = 1.$$

Let us introduce the following notation:

$$g(x) = limsup_{n \to \infty}|x_n| \qquad (x = (x_n)_{n \in \mathbf{N}} \in \mathbf{R}^{\mathbf{N}}),$$

$$G_\alpha = \{x \in \mathbf{R}^{\mathbf{N}} \; : \; g(x) = \alpha\} \qquad (\alpha \in [0, +\infty]).$$

Note that g may be regarded as an extended seminorm on the space $\mathbf{R}^{\mathbf{N}}$ (the phrase "extended" means here that the values of g run over the extended positive half-line). Also, it can easily be observed that g is Borel measurable. Analogous facts are valid for the functional

$$||x||_\infty = sup_{n \in \mathbf{N}}|x_n| \qquad (x = (x_n)_{n \in \mathbf{N}} \in \mathbf{R}^{\mathbf{N}})$$

which may be treated as an extended norm on $\mathbf{R}^{\mathbf{N}}$. As we know, the restrictions of the latter functional to the spaces c_0 and l_∞ coincide with usual norms on these spaces. In addition, the following relations are true:

$$g(x) \leq ||x||_\infty \qquad (x \in \mathbf{R}^{\mathbf{N}}),$$

$$g(x) = +\infty \Leftrightarrow ||x||_\infty = +\infty \qquad (x \in \mathbf{R}^{\mathbf{N}}).$$

Further, Lemma 4 shows that each set

$$G_{\alpha(\mu)} \qquad (\alpha(\mu) \in [0, +\infty])$$

is a support of a measure $\mu \in M_0$. Let us point out some other properties of these sets.

1. If $\alpha \neq \beta$, then $G_\alpha \cap G_\beta = \emptyset$.
2. $G_0 = c_0$.
3. $G_\alpha \subset l_\infty \setminus c_0 \Leftrightarrow \alpha \in]0, +\infty[$.
4. $G_{+\infty} = \{x \in \mathbf{R}^{\mathbf{N}} \; : \; ||x||_\infty = +\infty\}$.

Of course, the above-mentioned properties are very simple. Nevertheless, they imply some interesting consequences similar to the classical zero-one law. Indeed, it is clear that if $\mu \in M_0$, then each of the sets

$$c_0, \quad l_\infty \setminus c_0, \quad G_{+\infty}$$

is either of μ-measure zero or of μ-measure one. Simultaneously, we have the equality

$$\mu(c_0) + \mu(l_\infty \setminus c_0) + \mu(G_{+\infty}) = 1.$$

If $\alpha = 0$, then G_α is a vector subspace of the space $\mathbf{R}^{\mathbf{N}}$. If $\alpha \neq 0$, then G_α does not contain the neutral element of $\mathbf{R}^{\mathbf{N}}$, so it cannot be a vector subspace of this space.

In order to present further properties of sets G_α, it will be convenient to introduce the notation

$$U_r(y) = \{x \in \mathbf{R}^{\mathbf{N}} \ : \ ||x - y||_\infty < r\}$$

for any real number $r > 0$ and for any element $y \in \mathbf{R}^{\mathbf{N}}$. Note that if $y \in l_\infty$, then $U_r(y)$ coincides with the open ball in the Banach space l_∞, with centre y and radius r.

5. Let $0 < \alpha < +\infty$. Then, for all $r \in]0, \alpha[$, we have the relation

$$U_r(0) \cap G_\alpha = \emptyset.$$

If $\alpha = +\infty$, then the above-mentioned relation holds true for any $r > 0$.

6. Let $0 < \alpha < +\infty$. Then, for each $r > 0$, there exists an element $z \in l_\infty$ such that

$$U_r(z) \cap G_\alpha = \emptyset.$$

If $\alpha = +\infty$, then the above-mentioned relation is fulfilled for all $z \in l_\infty$.

The last two properties (5 and 6) can be proved without difficulty. We shall consider here only the case $0 < \alpha < +\infty$ (the case $\alpha = +\infty$ is left to the reader as an easy exercise).

In order to demonstrate 5, take any r from the interval $]0, \alpha[$ and suppose to the contrary that

$$U_r(0) \cap G_\alpha \neq \emptyset.$$

Then, choosing arbitrarily $x \in U_r(0) \cap G_\alpha$, we must have

$$\alpha = g(x) \leq ||x||_\infty < r \leq \alpha,$$

which yields a contradiction.

In order to demonstrate 6, we utilize the fact that g is an extended seminorm. Take any $r > 0$. Then there exists an element $z \in l_\infty$ such that

$$g(z) \geq r + \alpha.$$

Suppose to the contrary that

$$U_r(z) \cap G_\alpha \neq \emptyset$$

and pick an element x from this nonempty intersection. Then we get

$$\alpha = g(x) \geq g(z) - g(z - x) \geq g(z) - ||z - x||_\infty > \alpha,$$

which is impossible.

Lemma 4 and properties 5 and 6 of the sets G_α lead to the following proposition.

Lemma 5. *Let μ be a measure from M_0 and let $\mu(c_0) = 0$. Then, for each $r \in]0, \alpha(\mu)[$, the relation*

$$\mu(U_r(0)) = 0$$

is satisfied. Also, for any $r > 0$, there exists a vector $y \in l_\infty$ such that

$$\mu(U_r(y)) = 0.$$

Obviously, this lemma can be formulated in terms of random sequences. Namely, let $Y = (Y_n)_{n \in \mathbf{N}}$ be a random sequence from the class W_0, bounded with P-probability one, but not converging to zero with strictly positive P-probability. Then there always exists a real number $\rho > 0$ for which we have

$$P(\{sup_{n \in \mathbf{N}} |Y_n| < \rho\}) = 0,$$

and, for any real number $r > 0$, there exists an element $y = (y_n)_{n \in \mathbf{N}}$ of l_∞ such that

$$P(\{sup_{n \in \mathbf{N}} |Y_n - y_n| < r\}) = 0.$$

In particular, these two relations hold true for centered Gaussian random sequences (under the corresponding assumptions). It is reasonable to point out here that situations are possible where

$$P(\{sup_{n \in \mathbf{N}} |Y_n| = r\}) > 0$$

for some $r \in]0, +\infty[$ (in this connection, see e.g. Exercise 1). The effects of such type, for various classes of Gaussian sequences and Gaussian processes, were repeatedly mentioned in the literature (cf. [192]).

In the sequel, we need the following definition.

Definition 2. Let μ be a probability measure on $(\mathbf{R}^{\mathbf{N}}, B(\mathbf{R}^{\mathbf{N}}))$.

We say that μ is strongly locally scattered if there exists a real number $r > 0$ such that

$$\mu(U_r(y)) = 0$$

for all $y \in \mathbf{R}^{\mathbf{N}}$ or, equivalently, if

$$sup_{y \in \mathbf{R}^{\mathbf{N}}} \mu(U_r(y)) = 0.$$

We say that μ is scattered at zero if there exists a real number $r > 0$ for which

$$\mu(U_r(0)) = 0.$$

Finally, we say that μ is weakly locally dense if, for each $r > 0$, the relation

$$sup_{y \in \mathbf{R}^{\mathbf{N}}} \mu(U_r(y)) > 0$$

is valid.

We already know that if a measure μ from the class M_0 is not concentrated on the space c_0, then it turns out to be scattered at zero. For measures belonging to a certain subclass of M_0, we can even state that they are strongly locally scattered. Let us introduce the corresponding subclass of M_0.

Denote by M'_0 the family of all those measures from M_0 which are symmetric and unimodal. In other words, $\mu \in M'_0$ if and only if

$$\mu = \nu \circ A^{-1}$$

where $A \in V_0$ and ν is a product of a countable family of symmetric unimodal (in the sense of Khinchin) probability measures on $(\mathbf{R}, B(\mathbf{R}))$. We recall that all centered Gaussian measures from M_0 are in M'_0. So the investigation of the latter class is justified from the probabilistic viewpoint.

Also, it is reasonable to remark here that all measures of M'_0 satisfy the Anderson inequality (see Section 10). In particular, if $\mu \in M'_0$, then, for any real number $r > 0$ and for each vector $y \in \mathbf{R}^{\mathbf{N}}$, we have the relation

$$\mu(U_r(y)) \leq \mu(U_r(0)).$$

In virtue of this relation, we conclude that if $\mu \in M'_0$, then the following two assertions are equivalent:

(a) μ is scattered at zero;

(b) μ is strongly locally scattered.

This immediately implies the next statement.

Theorem 2. *Let μ be a measure from M'_0 and suppose that $\mu(c_0) = 0$. Then μ is strongly locally scattered.*

Now, we are able to specify the situation where the space c_0 is a support for measures from the class M'_0.

Theorem 3. *Let μ be a measure from M_0'. Then these two assertions are equivalent:*

1) $\mu(c_0) = 1$, i.e. c_0 is a support of μ;

2) μ is weakly locally dense.

Proof. Indeed, suppose that μ is weakly locally dense. In this case, μ cannot be strongly locally scattered and, according to Theorem 2, we must have $\mu(c_0) > 0$. Applying the zero-one law, we conclude that $\mu(c_0) = 1$.

Thus, the implication 2) $\Rightarrow$ 1) is established.

Conversely, suppose that $\mu(c_0) = 1$. Since c_0 is a separable Banach space with respect to the norm $||\cdot||_\infty$, we see that, for any real number $r > 0$, the inclusion

$$c_0 \subset \cup_{y\in Q} U_r(y)$$

holds true, where Q is an arbitrary countable dense subset of c_0. It follows from the inclusion above that

$$0 < 1 = \mu(c_0) \leq \sum_{y\in Q} \mu(U_r(y))$$

and, consequently, there exists at least one element $y \in c_0$ for which $\mu(U_r(y)) > 0$. This means that our μ is weakly locally dense.

Thus, the implication 1) $\Rightarrow$ 2) is valid.

Let us remark that the last step of the proof does not rely on the assumption $\mu \in M_0'$. Therefore, the implication 1) $\Rightarrow$ 2) is satisfied for any probability measure μ such that $\mu(c_0) > 0$. Note also that, for measures from the class M_0', this implication admits a much more stronger version, namely, if $\mu \in M_0'$ and $\mu(c_0) > 0$, then μ turns out to be dense at zero, i.e. for any $r > 0$, we have

$$\mu(U_r(0)) > 0.$$

Theorems 2 and 3 presented above enable us to formulate more precisely the thesis on the "essential infinite-dimensionality" (or even on the "essential nonseparability") of supports of those measures from M_0' which are concentrated on $l_\infty \setminus c_0$. In order to give here the precise formulation of this thesis, we first recall the definition of an outer measure. If μ is a probability measure on the space $(\mathbf{R}^\mathbf{N}, B(\mathbf{R}^\mathbf{N}))$ and D is an arbitrary subset of this space, then the μ-outer measure of D (denoted, as usual, by $\mu^*(D)$) is defined by the formula

$$\mu^*(D) = inf\{\mu(C) \ : \ C \in B(\mathbf{R}^\mathbf{N}), \ D \subset C\}.$$

Now, we can formulate the next statement.

Theorem 4. *Let $\mu \in M_0'$ and assume that $\mu(l_\infty \setminus c_0) = 1$. Then there exists no separable subset of l_∞ (with respect to the norm $||\cdot||_\infty$) having strictly positive outer μ-measure.*

Proof. Suppose to the contrary that there exists a separable set $D \subset l_\infty$ satisfying the relation

$$\mu^*(D) > 0.$$

Then, for any $r > 0$, this set can be covered by a countable family of balls of radius r, i.e.

$$D \subset \cup_{n \in \mathbf{N}} U_r(y_n)$$

where $\{y_n \ : \ n \in \mathbf{N}\}$ is some countable family of points in l_∞. Therefore, we get

$$0 < \mu^*(D) \leq \sum_{n \in \mathbf{N}} \mu(U_r(y_n))$$

whence it follows that our μ is weakly locally dense, which contradicts Theorem 2. The contradiction obtained finishes the proof.

Now, we are ready to discuss the question concerning the existence of those probability measures which are proper extensions of measures from the class M_0'. Here we deal again with the spaces c_0 and l_∞ which, as known, are Borel subsets of $\mathbf{R}^{\mathbf{N}}$ and, simultaneously, are Banach spaces with respect to the canonical norm $||\cdot||_\infty$. Let us recall that, since c_0 is separable, the σ-algebra $B_1(c_0)$ of cylindrical sets, induced by the σ-algebra $B(\mathbf{R}^{\mathbf{N}})$, coincides with the Borel σ-algebra $B(c_0)$. Consequently, if μ is a Borel probability measure on $\mathbf{R}^{\mathbf{N}}$ and $\mu(c_0) = 1$, then μ turns out to be a Borel probability measure on c_0 (and hence on l_∞).

For probability measures on l_∞, we have an essentially different situation, because the cylindrical σ-algebra $B_1(l_\infty)$ is smaller than the Borel σ-algebra $B(l_\infty)$. Therefore, if μ is a Borel probability measure on $\mathbf{R}^{\mathbf{N}}$ and $\mu(l_\infty) = 1$, then μ may be regarded as a probability measure only on $(l_\infty, B_1(l_\infty))$ and, in general, it is not clear whether μ can be extended to a Borel probability measure on l_∞. Note that the problem of the existence of an extension of a measure (given on $B_1(l_\infty)$) to a Borel measure on l_∞ is important for integration theory in infinite-dimensional vector spaces, because in many questions of this theory various Borel functions on l_∞ are usually considered and, consequently, appropriate Borel probability measures are needed for integration of those functions. The problem posed above has rather unexpected solution. Namely, it turns out that a measure μ from the class M_0' can be extended to a Borel probability measure on l_∞

if and only if $\mu(c_0) = 1$. We want to conclude this section of our book with a brief discussion of the result just formulated.

First, we wish to describe certain distinguishing features of those Gaussian measures from the class M_0' which admit an extension (with the property to be Gaussian) on some σ-algebra $B_2(l_\infty)$ satisfying the inclusions

$$B_1(l_\infty) \subset B_2(l_\infty) \subset B(l_\infty).$$

Lemma 6. *Let μ be a centered Gaussian measure from M_0' such that $\mu(l_\infty) = 1$ and let*

$$f : l_\infty \to \mathbf{R}$$

be a generalized Banach limit measurable with respect to μ. Then f turns out to be degenerate with respect to μ, i.e.

$$\mu(\{x \in l_\infty : f(x) = 0\}) = 1.$$

Proof. Suppose to the contrary that f is not degenerate. One of the properties of generalized Banach limits says that

$$|f(x)| \leq limsup_{n\to\infty}|x_n|$$

for all $x = (x_n)_{n\in\mathbf{N}} \in l_\infty$ (see Section 5). In view of the zero-one law, only two cases are possible for μ: either $\mu(c_0) = 1$ or $\mu(l_\infty \setminus c_0) = 1$. If $\mu(c_0) = 1$, the assertion is trivial because the restriction of any generalized Banach limit to c_0 is identically zero. Suppose now that

$$\mu(l_\infty \setminus c_0) = 1.$$

Then, taking into account the linearity and measurability of f, we infer that f is a centered Gaussian random variable on some probability space $(l_\infty, B_2(l_\infty), \mu)$, with variance

$$\sigma_f^2 = \int_{l_\infty} f^2(x)\mu(dx) > 0.$$

Further, according to Lemma 4, there exists a real number $\alpha \in]0, +\infty[$ such that

$$\mu(\{x \in l_\infty : |f(x)| \leq \alpha\}) = 1.$$

Obviously, this equality can be rewritten in terms of the distribution density of f, and we obtain the relation

$$\frac{1}{(2\pi)^{1/2}\sigma_f}\int_{-\alpha}^{\alpha} exp(-\frac{t^2}{2\sigma_f^2})dt = 1.$$

But the latter equality is impossible since α is finite. The contradiction obtained ends the proof of Lemma 6.

Thus, we conclude that if a centered Gaussian measure belonging to the class M_0' and concentrated on $l_\infty \setminus c_0$ admits a Gaussian extension, then each generalized Banach limit is either nonmeasurable with respect to the extension, or vanishes almost everywhere.

Our further considerations need some more deep information about supports of Borel probability measures on metric spaces. Let us recall Theorem 1 from Section 8, stating that if a Borel probability measure μ is given on a metric space whose topological weight is not real-valued measurable (see Appendix 3), then there exists a separable support of μ. We are going to apply this general theorem in a concrete situation. Namely, as we know, the topological weight of l_∞ is equal to the cardinality continuum. Therefore, if we assume that the cardinality continuum is not real-valued measurable, then we may utilize the above-mentioned theorem to Borel probability measures on l_∞.

Theorem 5. *Let $\mu \in M_0'$ and $\mu(l_\infty) = 1$. Suppose also that the cardinality continuum is not real-valued measurable. Then the following two assertions are equivalent:*

1) $\mu(c_0) = 1$;

2) μ can be extended to a Borel probability measure on l_∞.

Proof. Indeed, if $\mu(c_0) = 1$, then, as mentioned earlier, μ may be regarded as a Borel probability measure on the space c_0 and hence on the space l_∞.

Conversely, suppose that there exists an extension μ' of μ defined on the whole Borel σ-algebra of l_∞. Then, in accordance with our assumption, there is a closed separable subset D of l_∞ such that

$$\mu'(D) = 1.$$

Now, applying to D the same argument as in the proof of Theorem 4, we infer that, for any real number $r > 0$, there exists a vector $y \in l_\infty$ satisfying the relation

$$\mu'(U_r(y)) > 0.$$

But we have

$$U_r(y) \in B_1(l_\infty)$$

and, since μ' is an extension of μ, we may write

$$\mu(U_r(y)) = \mu'(U_r(y)) > 0,$$

which yields that our μ is weakly locally dense. A direct application of Theorem 3 ends the proof.

The next statement is an immediate consequence of the result just obtained.

Theorem 6. *Let $\mu \in M_0'$ and suppose that*

$$\mu(l_\infty) = 1, \quad \mu(c_0) = 0.$$

Then μ cannot be extended to a Borel probability measure on l_∞.

Roughly speaking, Theorem 6 shows that there are no Borel measures in the class M_0' concentrated on the set $l_\infty \setminus c_0$.

Summarizing the two last results, let us stress again that, for a measure μ from M_0', the question whether $\mu(c_0) = 1$ is equivalent to the problem of the existence of a Borel extension of μ. This fact demonstrates the difficulty of the problem of characterization of those measures from M_0' which are concentrated on c_0 and, simultaneously, outlines a new approach to this problem. Besides, it turns out that if a measure belonging to M_0' is not concentrated on c_0, then it cannot be Borel on l_∞. In this connection, we wish to emphasize once more that the class M_0' is very natural for probability theory and for integration theory in infinite-dimensional topological vector spaces.

So, as shown above, a suitable combination of the Anderson inequality with the theorem on the existence of separable supports of Borel probability measures (given in metric spaces with nonmeasurable topological weights) enables to obtain some deep information about oscillation properties of random sequences.

In connection with the material presented in this section, we refer the reader to [37], [42], [43] and [45].

For oscillation properties of trajectories of Gaussian stochastic processes, see [97] and [192] (see also the next section of the book, where it is demonstrated that the Anderson inequality plays an essential role in study of the behaviour of trajectories of Gaussian processes).

EXERCISES

1. Let $(\sigma_n)_{n\in\mathbf{N}}$ be a sequence of strictly positive real numbers, such that

$$lim_{n\to\infty}\sigma_n = 0,$$

and let $X = (X_n)_{n\in\mathbf{N}}$ be a sequence of independent $N(0,\sigma_n^2)$-distributed Gaussian random variables.

a) Prove that $\alpha(X) = \gamma(X)$ where

$$\gamma(X) = inf\{\varepsilon > 0 \ : \ \sum_{n\in\mathbf{N}} \sigma_n exp(-\frac{\varepsilon}{2\sigma_n^2}) < +\infty\}.$$

b) Show that if the series

$$\sum_{n\in\mathbf{N}} \sigma_n exp(-\frac{\gamma(X)}{2\sigma_n^2})$$

is convergent, then

$$P(\{sup_{n\in\mathbf{N}}|X_n| = \gamma(X)\}) > 0.$$

c) Check that the series of b) is convergent for

$$\sigma_n^2 = \frac{1}{2(ln(n) + ln(ln(n)))} \qquad (n \geq 3).$$

In this case, we have $\gamma(X) = 1$.

2. Let X be a sequence of nondegenerate independent symmetric random variables and let A be an infinite triangular matrix. Put $Y = AX$. Prove that

$$P(\{Y \in c_0\}) = 1 \Rightarrow A \in V_0.$$

3. Let $Y = (Y_n)_{n\in\mathbf{N}}$ be a centered Gaussian sequence from the class W_0. Demonstrate that

$$P(\{Y \in l_\infty\}) = 1 \Rightarrow lim_{n\to\infty}E(Y_n^2) = 0.$$

4. Let $(Y_n)_{n\in\mathbf{N}}$ be a standard Gaussian sequence. Show that, for P-almost all $\omega \in \Omega$, the equality

$$D((Y_n(\omega))_{n\in\mathbf{N}}) = [-\infty, +\infty]$$

holds true.

5. Give an example of a sequence $(Y_n)_{n \in \mathbf{N}}$ of independent centered Gaussian random variables, such that

$$D((Y_n(\omega))_{n \in \mathbf{N}}) = \{-\infty,\ 0,\ +\infty\}$$

for P-almost all $\omega \in \Omega$.

6. Let μ be a centered Gaussian measure on the space $(\mathbf{R}^{\mathbf{N}}, B(\mathbf{R}^{\mathbf{N}}))$ and let F be a μ-measurable vector subspace of $\mathbf{R}^{\mathbf{N}}$. Show that either $\mu(F) = 0$ or $\mu(F) = 1$.

7. Generalize Theorems 2 - 6 from this section in the following manner. Take an arbitrary separable Banach space F instead of c_0, consider the space F^{**} instead of $l_\infty = (c_0)^{**}$ and, finally, take a centered Gaussian measure μ in F^{**} defined on the cylindrical σ-algebra $B(F^{**}, F^*)$.

Examine also the situation when some centered Gaussian premeasure ν is given on the cylindrical algebra $S(F^{**}, F^*)$.

12. Comparison principles for Gaussian processes

Here we demonstrate how the study of Gaussian stochastic processes is possible by comparing their covariance functions. For this purpose, we need some preordering relations in the class $\Phi_+(T)$ of real-valued positive functions given on a set T of parameters. As a rule, T will be a separable metric space. In particular, the following situations are of especial interest to us:

$T = \{1, 2, ..., m\}$ where $m \geq 1$ is a natural number;

$T = \mathbf{N}$;

$T = [0, 1]$;

$T = \mathbf{R}$ or $T = \mathbf{R}_+ = \{t \in \mathbf{R} \ : \ t \geq 0\}$.

As mentioned in Section 10, the class $\Phi_+(T)$ coincides with the family of covariance functions for centered Gaussian processes given on a set T of parameters. Therefore, any preordering in the class $\Phi_+(T)$ may be regarded as a preordering in the class $\Gamma(T)$ of all those separable centered Gaussian processes $X = (X_t)_{t \in T}$ which are defined on some fixed probability space (Ω, S, P), or as a preordering in the class $M(\mathbf{R}^T)$ of all those centered Gaussian probability measures which are defined on the space

$$(\mathbf{R}^T, B(\mathbf{R}^T, (\mathbf{R}^T)^*)).$$

The preorderings of this sort will be called covariance preorderings. The necessity of consideration of such preorderings is dictated by the circumstance that there are many nonlinear functionals on the class of Gaussian processes, whose distributions do not admit a visual representation. This can be said, for example, about the functionals:

$$sup_{t \in T} X_t, \qquad sup_{t \in T} |X_t|,$$

which are important from the purely theoretical viewpoint and play a significant role in various applications. Since the distributions of Gaussian processes are completely determined by their covariance functions, it is natural to anticipate that a suitable choice of a covariance preordering leads to the corresponding stochastic preordering between processes. This correspondence can be expressed by different kinds of inequalities which are closely connected with the convexity structure. We shall see below that the Anderson inequality turns out to be a technical tool for obtaining some useful results concerning comparison of Gaussian processes. Moreover, it will be clear in several cases below that the unimodality of a Gaussian probability measure is not essential, and the corresponding inequalities can be established by starting with the notion of a symmetric probability measure. However, the usage of unimodality (i.e. of the Anderson inequality) yields the possibility to get much more precise results. The comparison method described in this section enables us to study efficiently the properties of realizations of Gaussian processes and, in addition, the method demonstrates once more close relationships between the oscillation properties of trajectories and the problem of extending a measure (see the preceding section). Note that this topic still remains under extensive investigations.

Let us take any two functions

$$K_1 = (K^{(1)}_{t,s})_{t\in T,\ s\in T}, \quad K_2 = (K^{(2)}_{t,s})_{t\in T,\ s\in T}$$

from the class $\Phi_+(T)$. For these functions, define

$$K_2 \leq_\zeta K_1 \Leftrightarrow K_1 - K_2 \in \Phi_+(T).$$

Then it is easy to see that $\leq_\zeta$ is an ordering relation in $\Phi_+(T)$.

Further, put

$$K_2 \leq_\eta K_1$$

if and only if

$$K^{(1)}_{t,s} \leq K^{(2)}_{t,s} \qquad (t \in T,\ s \in T),$$

$$K^{(1)}_{t,t} = K^{(2)}_{t,t} \qquad (t \in T).$$

It is not hard to check that $\leq_\eta$ is also an ordering relation in $\Phi_+(T)$.

Finally, put

$$K_2 \leq_\xi K_1$$

if and only if

$$\sigma^{(2)}_{t,s} \leq \sigma^{(1)}_{t,s} \qquad (t \in T,\ s \in T)$$

where

$$\sigma^{(j)}_{t,s} = K^{(j)}_{t,t} - 2K^{(j)}_{t,s} + K^{(j)}_{s,s} \qquad (j = 1, 2).$$

It should be noted that $\leq_\xi$ is a preordering relation in $\Phi_+(T)$. Moreover, it can easily be checked that

$$K_2 \leq_\zeta K_1 \Rightarrow K_2 \leq_\xi K_1,$$

$$K_2 \leq_\eta K_1 \Rightarrow K_2 \leq_\xi K_1.$$

The above-mentioned preorderings $\leq_\zeta$, $\leq_\eta$ and $\leq_\xi$ will be called covariance preorderings in the class $\Gamma(T)$ of Gaussian processes or, equivalently, covariance preorderings in the class $M(\mathbf{R}^T)$ of Gaussian measures (let us stress once more that the first two preorderings are, in fact, orderings). In the sequel, it will be convenient to speak of the covariance preorderings in terms of stochastic processes and in terms of probability measures. For instance, if X and Y are any two processes from the class $\Gamma(T)$, with covariance functions K_X and K_Y respectively, then we define

$$Y \leq X \Leftrightarrow K_Y \leq K_X$$

where $\leq$ is any of the above-mentioned preorderings. Analogously, if μ_1 and μ_2 are any two measures from the class $M(\mathbf{R}^T)$, then we put

$$\mu_2 \leq \mu_1 \Leftrightarrow K_2 \leq K_1$$

where K_j denotes the covariance of μ_j $(j = 1, 2)$.

Let us give a probabilistic interpretation of the ordering $\leq_\zeta$. If some functions K_1 and K_2 belong to the class $\Phi_+(T)$, then they are covariance functions of the corresponding centered Gaussian processes

$$X_1 = (X_t^{(1)})_{t\in T}, \qquad (X_t^{(2)})_{t\in T}$$

which are defined on a probability space (Ω, S, P). Further, if we have $K_2 \leq_\zeta K_1$, then it is sometimes convenient to realize this probability space as the product of three spaces:

$$\prod_{1\leq j\leq 3} (\mathbf{R}^T, B(\mathbf{R}^T, (\mathbf{R}^T)^*), \mu_j),$$

where μ_1 and μ_2 are Gaussian measures on $\mathbf{R}^T$ with covariance functions K_1 and K_2 respectively, and μ_3 is a probability measure on $\mathbf{R}^T$ described as follows. Since $K_2 \leq_\zeta K_1$, the difference $K_3 = K_1 - K_2$ belongs to $\Phi_+(T)$, too, and we define μ_3 as the centered Gaussian measure on $\mathbf{R}^T$ with covariance function K_3. We denote by $X_3 = (X_t^{(3)})_{t\in T}$ the corresponding centered Gaussian process whose covariance function coincides with K_3. In view of the choice of our basic probability space (Ω, S, P), the processes X_1, X_2, X_3 can be taken in such a way that they become mutually independent. This implies the relation

$$cov(X_2 + X_3) = cov(X_2) + cov(X_3) =$$

$$K_2 + (K_1 - K_2) = K_1 = cov(X_1)$$

whence it immediately follows that the process X_1 is equidistributed with the sum $X_2 + X_3$. We thus see that the following auxiliary proposition holds true.

Lemma 1. *Let X and Y be any two processes from the class $\Gamma(T)$, such that $Y \leq_\varsigma X$. Then there exist a probability space and three centered Gaussian processes X_1, Y_1, Z_1 on it, such that X_1 and Y_1 are some copies of X and Y respectively, the processes Y_1 and Z_1 are independent, and*

$$X_1 = Y_1 + Z_1.$$

This proposition admits an equivalent formulation in terms of Gaussian measures:

Lemma 2. *Let μ_1 and μ_2 be any two measures from the class $M(\mathbf{R}^T)$, such that $\mu_2 \leq_\varsigma \mu_1$. Then there exists a measure μ_3 from the same class, for which we have*

$$\mu_1 = \mu_2 * \mu_3$$

where the symbol $*$ denotes the convolution operation for measures.

We know that all centered Gaussian measures on the space $\mathbf{R}^\mathbf{N}$ are symmetric and unimodal in the sense of Anderson (see Section 10). In our further considerations, we need another version of the Anderson inequality. This version is formulated in the next statement.

Lemma 3. *Let $F = \mathbf{R}^\mathbf{N}$, let $\mu_2 \in M(F)$ and let μ_3 be an arbitrary Borel probability measure on F. Denote*

$$\mu_1 = \mu_2 * \mu_3.$$

Then, for every convex symmetric (with respect to zero) Borel subset A of F, we have the inequality

$$\mu_1(A) \leq \mu_2(A).$$

Proof. In virtue of the Anderson inequality, we may write

$$\mu_2(A - z) \leq \mu_2(A)$$

for any vector $z \in F$. Now, applying the fact that μ_1 is the convolution of μ_2 and μ_3, we obtain

$$\mu_1(A) = \int_F \mu_2(A - z)\mu_3(dz) \leq \int_F \mu_2(A)\mu_3(dz) = \mu_2(A).$$

This completes the proof.

Lemmas 1 and 3 imply at once the following comparison theorem for the ordering $\leq_\zeta$.

Theorem 1. *Let T be an arbitrary separable metric space and let $X = (X_t)_{t\in T}$ and $Y = (Y_t)_{t\in T}$ be any two processes from the class $\Gamma(T)$, such that $Y \leq_\zeta X$. Then, for each real number $r > 0$, the inequality*

$$P(\{sup_{t\in T}|Y_t| > r\}) \leq P(\{sup_{t\in T}|X_t| > r\})$$

holds true.

Proof. Indeed, if the given space T is at most countable, then this assertion is a direct consequence of Lemmas 1 and 3. The case when $card(T) > card(\mathbf{N})$ can be reduced to the previous one, since all processes from the class $\Gamma(T)$ are assumed to be separable (so it suffices to consider the union of two countable subsets of T, the first of which is a set of separability for X and the second one is a set of separability for Y). These trivial remarks end the proof of the theorem.

As an illustration, the following typical example is relevant.

Example 1. Let X and Y be any two processes from the class $\Gamma(\mathbf{N})$ satisfying the relation $Y \leq_\zeta X$. Then the implications

$$P(\{X \in l_\infty\}) = 1 \Rightarrow P(\{Y \in l_\infty\}) = 1,$$

$$P(\{X \in c_0\}) = 1 \Rightarrow P(\{Y \in c_0\}) = 1$$

are valid. Here the symbols l_∞ and c_0 have their usual meaning.

The above-mentioned implications follow directly from Theorem 1. In the sequel, we shall demonstrate that they remain true under weaker assumptions. The next statement seems to be much more interesting.

Theorem 2. *Let F be a separable Banach space (continuously embedded in $\mathbf{R}^{\mathbf{N}}$ with the aid of a countable subset of F^* which separates the points of F) and let μ_1 and μ_2 be two centered Gaussian cylindrical premeasures given on the algebra $S(F, F^*)$, with covariance functions $K_1 \in \Phi_+(\mathbf{N})$ and $K_2 \in \Phi_+(\mathbf{N})$ respectively. Suppose that $K_2 \leq_\zeta K_1$ and that μ_1 can be extended to a Borel probability measure on F. Then μ_2 can also be extended to a Borel probability measure on F.*

Proof. In virtue of the general criterion (see Section 6), a cylindrical probability premeasure μ defined on $S(F, F^*)$ admits an extension to a

Borel probability measure on F if and only if, for any $\varepsilon > 0$, there exists a compact set $B_\varepsilon \subset F$ such that

$$(1) \quad \mu(C) \leq \varepsilon \qquad (C \in S(F, F^*),\ C \cap B_\varepsilon = \emptyset).$$

We may assume, without loss of generality, that B_ε is convex and symmetric (otherwise, it suffices to consider the closed convex symmetric hull of B_ε which is compact, as well). Further, according to a direct consequence of the Hahn-Banach theorem, we may write

$$B_\varepsilon = \{x \in F \ :\ sup_{n \in \mathbf{N}} |g_n^{(\varepsilon)}(x)| \leq 1\}$$

for some countable family $(g_n^{(\varepsilon)})_{n \in \mathbf{N}} \subset F^*$ (depending on B_ε). Now, it is clear that relation (1) holds true for all $\varepsilon > 0$ if and only if

$$lim_{m \to \infty} \mu(\{x \in F \ :\ sup_{n \leq m} |g_n^{(\varepsilon)}(x)| > 1\}) \leq \varepsilon.$$

But, applying Lemma 3, we obtain

$$lim_{m \to \infty} \mu_2(\{x \in F \ :\ sup_{n \leq m} |g_n^{(\varepsilon)}(x)| > 1\}) \leq$$

$$lim_{m \to \infty} \mu_1(\{x \in F \ :\ sup_{n \leq m} |g_n^{(\varepsilon)}(x)| > 1\}).$$

This shows us that μ_2 can be extended to a Borel probability measure on F, and the proof is completed.

The ordering $\leq_\zeta$ is rather strong and is used relatively rarely. However, this ordering turns out to be very helpful in combination with the ordering $\leq_\eta$ which is more interesting for us. It should be noted that these two orderings can be fulfilled (simultaneously) only in trivial cases (cf. Exercise 3). Therefore, the ordering $\leq_\eta$ takes into account some additional information about relationships between the covariance function and the distribution of a Gaussian process. Let us also recall that $\leq_\eta$ is defined in the class $\Gamma(T)$ as follows:

$$(Y_t)_{t \in T} \leq_\eta (X_t)_{t \in T}$$

holds if and only if

$$E(X_t X_s) \leq E(Y_t Y_s) \qquad (t \in T,\ s \in T),$$

$$E(X_t^2) = E(Y_t^2) \qquad (t \in T).$$

Thus, the relation $Y \leq_\eta X$ implies that the correlation (i.e. normalized covariance) of X is less than or equal to the correlation of Y.

In connection with the ordering $\leq_\eta$, it is reasonable to present here the Slepian inequality which plays a significant role in the theory of Gaussian processes. Namely, let us formulate and prove the following statement due to Slepian (see [184]).

Theorem 3. *Let T be a separable metric space, let $X = (X_t)_{t\in T}$ and $Y = (Y_t)_{t\in T}$ be two processes from the class $\Gamma(T)$, and suppose that*

$$Y \leq_\eta X.$$

Then, for all $r \in \mathbf{R}$, we have the inequality

$$P(\{sup_{t\in T} Y_t > r\}) \leq P(\{sup_{t\in T} X_t > r\}).$$

Proof. We may concentrate our attention only on the case when T is a finite set with $card(T) = m$. Indeed, the case when $T = \mathbf{N}$ can be deduced from the first one by tending m to infinity. Finally, the case of an arbitrary separable metric space T is reduced to the second one by applying the separability of given processes X and Y.

So, in the sequel, we assume that

$$T = \{1, 2, ..., m\} \qquad (m > 1).$$

Let μ be a centered Gaussian measure on the space $\mathbf{R}^m$, whose covariance matrix

$$K = (K_{jn})_{1\leq j\leq m,\ 1\leq n\leq m}$$

is strictly positive. First of all, we need a certain system of differential equations satisfied by the density p of μ. For the sake of convenience, we shall write

$$p(z) = p(z, K) = p(z_1, z_2, ..., z_m, K)$$

where $z = (z_1, z_2, ..., z_m) \in \mathbf{R}^m$. Recall now that the characteristic functional

$$\phi_\mu(x) = \int exp(i < x, z >)\mu(dz) \qquad (x \in \mathbf{R}^m)$$

of our Gaussian measure μ has the form

$$\phi_\mu(x) = exp(-(1/2) < Kx, x >).$$

Applying the well-known inversion formula for characteristic functionals (see, e.g., [74], [124] or [180]), we get

$$p(z, K) = (2\pi)^{-m} \int exp(i < x, z >)\phi_\mu(x)dx =$$

$$(2\pi)^{-m} \int\int \dots \int exp(i \sum_{1 \le n \le m} x_n z_n -$$

$$(1/2) \sum_{1 \le j \le m,\ 1 \le n \le m} K_{jn} x_j x_n) dx_1 dx_2 \dots dx_m.$$

Starting with this relation and taking the corresponding partial derivatives, we easily come to the system of differential equations

$$\frac{\partial}{\partial K_{jn}} p(z, K) = \frac{\partial^2}{\partial z_j \partial z_n} p(z, K) \qquad (j < n).$$

Here the equalities are written only for $j < n$, because we have

$$K_{jn} = K_{nj} \qquad (1 \le j \le m,\ 1 \le n \le m).$$

Further, for any real number u, we denote

$$\Pi = \Pi(u) = \{z \in \mathbf{R}^m \ :\ sup_{1 \le n \le m} z_n \le u\}.$$

Then it is clear that the relation

$$\mu(\Pi) = \mu(\Pi, K) = \int_{-\infty}^{u} \int_{-\infty}^{u} \dots \int_{-\infty}^{u} p(z_1, z_2, \dots, z_m, K) dz_1 dz_2 \dots dz_m$$

holds true. Combining this relation with the above-mentioned equations, we easily obtain

$$\frac{\partial}{\partial K_{jn}} \mu(\Pi, K) =$$

$$\int_{-\infty}^{u} \int_{-\infty}^{u} \dots \int_{-\infty}^{u} \frac{\partial^2}{\partial z_j \partial z_n} p(z_1, z_2, \dots, z_m, K) dz_1 dz_2 \dots dz_m$$

for $j < n$. Integrating the last expression with respect to z_j and z_n, we see that $\frac{\partial}{\partial K_{jn}} \mu(\Pi, K)$ is equal to

$$\int_{-\infty}^{u} \dots \int_{-\infty}^{u} p(z_1, \dots, z_{j-1}, u, z_{j+1}, \dots, z_{n-1}, u, z_{n+1}, \dots, z_m, K)$$

$$dz_1 \dots dz_{j-1} dz_{j+1} \dots dz_{n-1} dz_{n+1} \dots dz_m.$$

Now, taking into account the fact that

$$p(z, K) > 0 \qquad (z \in \mathbf{R}^m),$$

we conclude that

$$(1) \quad \frac{\partial}{\partial K_{jn}}\mu(\Pi, K) > 0 \qquad (j < n).$$

Formula (1) will be basic for our further considerations.

Let μ_X and μ_Y denote the distributions of X and Y respectively. These distributions are some centered Gaussian measures on $\mathbf{R}^m$ with the corresponding covariance matrices

$$K_X = (K_{jn}^{(1)})_{1\leq j\leq m,\ 1\leq n\leq m}, \qquad K_Y = (K_{jn}^{(2)})_{1\leq j\leq m,\ 1\leq n\leq m}.$$

In virtue of our assumption $Y \leq_\eta X$, we have

$$K_{jn}^{(1)} \leq K_{jn}^{(2)} \qquad (1 \leq j \leq m,\ 1 \leq n \leq m),$$

$$K_{nn}^{(1)} = K_{nn}^{(2)} \qquad (1 \leq n \leq m).$$

Suppose first that the matrices K_X and K_Y are strictly positive. Then, for each number $r \in [0, 1]$, the matrix

$$K(r) = rK_Y + (1 - r)K_X$$

is strictly positive, too. Therefore, $K(r)$ turns out to be the covariance matrix of some centered Gaussian measure $\mu = \mu(\cdot, K(r))$ on $\mathbf{R}^m$, and it is clear that an analogue of relation (1) holds true for μ. So we may write

$$\frac{d}{dr}\mu(\Pi, K(r)) = \sum_{1\leq j\leq n\leq m} (\frac{\partial}{\partial K_{jn}}\mu(\Pi, K(r)))\frac{dK_{jn}(r)}{dr} =$$

$$\sum_{1\leq n\leq m} (\frac{\partial}{\partial K_{nn}}\mu(\Pi, K(r)))(K_{nn}^{(2)} - K_{nn}^{(1)}) +$$

$$\sum_{j<n}(\frac{\partial}{\partial K_{jn}}\mu(\Pi, K(r)))(K_{jn}^{(2)} - K_{jn}^{(1)}) \geq 0.$$

In other words, $\mu(\Pi, K(r))$ is increasing with respect to r. From this fact we infer, for any $u \in \mathbf{R}$, that

$$P(\{sup_{1\leq n\leq m} X_n \leq u\}) = \mu(\Pi(u), K(0)) \leq$$

$$\mu(\Pi(u), K(1)) = P(\{sup_{1\leq n\leq m} Y_n \leq u\}),$$

which yields the desired inequality in the case of strictly positive K_X and K_Y. Thus, it remains to free of the assumption that K_X and K_Y are strictly positive. For this purpose, fix a number $\varepsilon > 0$ and consider the Gaussian random vectors

$$X_\varepsilon = X + \varepsilon Z, \quad Y_\varepsilon = Y + \varepsilon Z,$$

where Z is a centered Gaussian vector in $\mathbf{R}^m$ independent with respect to X and Y and having the unity covariance matrix $I = I_Z$. Then we get

$$K_{X_\varepsilon} = K_X + \varepsilon I, \quad K_{Y_\varepsilon} = K_Y + \varepsilon I,$$

and the matrices K_{X_ε} and K_{Y_ε} turn out to be strictly positive. In addition,

$$K_{Y_\varepsilon} \leq_\eta K_{X_\varepsilon}.$$

Consequently, as shown above, the Slepian inequality is valid for X_ε and Y_ε. Now, tending ε to zero, we conclude that the same inequality must be satisfied for the original random vectors X and Y. This completes the proof of Theorem 3.

Now, we are able to study some properties of the preordering $\leq_\xi$. As mentioned earlier, $\leq_\xi$ is the weakest preordering among all preorderings introduced in this section. Therefore, the statements established for $\leq_\xi$ will be true for $\leq_\eta$ and $\leq_\zeta$ as well. Note that if two centered Gaussian processes $X = (X_t)_{t\in T}$ and $Y = (Y_t)_{t\in T}$ have the same variances, i.e.

$$E(X_t^2) = E(Y_t^2) \quad (t \in T),$$

then the preorderings $\leq_\eta$ and $\leq_\xi$ are equivalent for X and Y. Also, dealing with processes from the class $\Gamma(T)$, we may define $\leq_\xi$ in another (equivalent) form. Namely, we see that $Y \leq_\xi X$ if and only if

$$E((Y_t - Y_s)^2) \leq E((X_t - X_s)^2) \quad (t \in T, \ s \in T).$$

Let us give a simple example showing that an analogue of the Slepian inequality does not hold for $\leq_\xi$.

Example 2. Let $X = (X_n)_{n\in\mathbf{N}}$ be a centered Gaussian random sequence and let Z_0 be a Gaussian $N(0,1)$-distributed random variable independent with respect to X. Fix real numbers $\sigma > 0$, $\varepsilon \in]0,1[$ and define a random sequence $Y = (Y_n)_{n\in\mathbf{N}}$ by putting

$$Y_n = \varepsilon X_n + \sigma Z_0 \quad (n \in \mathbf{N}).$$

Then, for all t and s from $\mathbf{N}$, we have

$$E((Y_t - Y_s)^2) = \varepsilon^2 E((X_t - X_s)^2) < E((X_t - X_s)^2)$$

whence it follows that $Y \leq_\xi X$. At the same time, for any $r > 0$ and for sufficiently large σ, the relation

$$P(\{sup_{n\in\mathbf{N}} X_n > r\}) < P(\{sup_{n\in\mathbf{N}}(\varepsilon X_n + \sigma Z_0) > r\})$$

is fulfilled, which contradicts the Slepian inequality.

Nevertheless, a suitable analogue of Theorems 1 and 3 can be obtained for $\leq_\xi$. We only have to compare the process Y with a new process obtained by adding to X some independent random variable. This idea enables us to utilize simultaneously the Slepian inequality (Theorem 3) and the Anderson inequality (Theorem 1).

Let us formulate and prove the corresponding result.

Theorem 4. *Let T be a separable metric space, let $X = (X_t)_{t\in T}$ and $Y = (Y_t)_{t\in T}$ be processes from the class $\Gamma(T)$ and suppose that*

$$Y \leq_\xi X.$$

Then, for each real number $r > 0$, we have

$$P(\{sup_{t\in T}|Y_t| > r\}) \leq 2P(\{sup_{t\in T}(X_t + b_t Z_0) > r\})$$

where Z_0 is a Gaussian $N(0,1)$-distributed random variable independent with respect to X, and

$$b_t^2 = sup_{t\in T}(max(0, E(X_t^2) - E(Y_t^2))) - (E(X_t^2) - E(Y_t^2)) \quad (t \in T).$$

Proof. Define $h \geq 0$ by the equality

$$h^2 = sup_{t\in T}(max(0, E(X_t^2) - E(Y_t^2)))$$

and consider two new processes $X' = (X'_t)_{t\in T}$ and $Y' = (Y'_t)_{t\in T}$ from the class $\Gamma(T)$, given by the formulae

$$X'_t = X_t + b_t Z_0 \quad (t \in T),$$

$$Y'_t = Y_t + hZ_0 \quad (t \in T),$$

where a Gaussian $N(0,1)$-distributed random variable Z_0 is chosen to be independent with respect to X and Y. Then a direct verification shows that

$$E((Y'_t)^2) = E((X'_t)^2) \quad (t \in T).$$

Further, since $Y \le_\xi X$, we have

$$E((Y'_t - Y'_s)^2) = E((Y_t - Y_s)^2) \le E((X_t - X_s)^2) \le$$

$$E((X_t - X_s)^2) + (b_t - b_s)^2 = E((X'_t - X'_s)^2)$$

for all t and s from T. This means that $Y' \le_\eta X'$. Hence, according to Theorem 3,

$$P(\{sup_{t\in T} Y'_t > r\}) \le P(\{sup_{t\in T} X'_t > r\})$$

for any $r \in \mathbf{R}$. Moreover, taking into account the definition of the process Y', we see that $Y \le_\zeta Y'$. Now, applying Theorem 1, we deduce that

$$P(\{sup_{t\in T}|Y_t| > r\}) \le P(\{sup_{t\in T}|Y'_t| > r\}).$$

Finally, utilizing the fact that the process Y' is symmetric, we come to the relation

$$P(\{sup_{t\in T}|Y_t| > r\}) \le P(\{sup_{t\in T}|Y'_t| > r\}) \le 2P(\{sup_{t\in T} Y'_t > r\}) \le$$

$$2P(\{sup_{t\in T} X'_t > r\}) = 2P(\{sup_{t\in T}(X_t + b_t Z_0) > r\}),$$

which implies the assertion of Theorem 4.

Let us remark that if processes X and Y belong to $\Gamma(T)$ and $Y \le_\eta X$ (hence $Y \le_\xi X$), then $b_t = 0$ for all $t \in T$, and the inequality of Theorem 4 is reduced to the following simplified formula:

$$P(\{sup_{t\in T}|Y_t| > r\}) \le 2P(\{sup_{t\in T} X_t > r\}) \qquad (r > 0),$$

which is completely compatible with the Slepian inequality.

Some additional information concerning the above-mentioned results is presented in exercises to this section. Here we only wish to point out one consequence of Theorem 4, obtained in [39], which turns out to be helpful (as a technical tool) in our further considerations.

Lemma 4. *Let T be a separable metric space, let X and Y be two processes from the class $\Gamma(T)$ and suppose that*

$$Y \le_\xi X.$$

Then, for each real number $r > 0$, we have

$$P(\{sup_{t\in T}|Y_t| > 2r\}) \le 2P(\{sup_{t\in T}|X_t| > r\}) + 4exp(-r^2/\Delta(T))$$

where

$$\Delta(T) = 4sup_{t\in T}|E(X_t^2) - E(Y_t^2)|.$$

Proof. Indeed, using the notation of Theorem 4, we may write

$$P(\{sup_{t\in T}(X_t + b_t Z_0) > 2r\}) \leq P(\{sup_{t\in T}|X_t| > r\}) +$$

$$P(\{(sup_{t\in T} b_t)|Z_0| > r\}),$$

$$P(\{(sup_{t\in T} b_t)|Z_0| > r\}) \leq P(\{(\Delta(T)/2)^{1/2}|Z_0| > r\}) \leq 2exp(-r^2/\Delta(T)).$$

In view of Theorem 4, these relations immediately imply the required result.

Now, we are going to discuss the following natural question. Suppose that X and Y are processes from $\Gamma(T)$ and $Y \leq_\xi X$. Assume also that P-almost all trajectories of the process X are regular in some sense (e.g. are bounded, continuous, etc.). Can we assert that P-almost all trajectories of the process Y are regular in the same sense? This question is important not only for the theory of stochastic processes. It is rather interesting from the viewpoint of the theory of measures in infinite-dimensional function spaces. Indeed, the question admits a formulation in terms of covariance functions. More precisely, let K_1 and K_2 be two functions from the class $\Phi_+(T)$ and suppose that $K_2 \leq_\xi K_1$. Suppose also that K_1 is the covariance function of a Gaussian measure given on some vector space $F \subset \mathbf{R}^T$. Does there exist, on the same space F, a Gaussian measure whose covariance function coincides with K_2?

We shall demonstrate below that, in many cases, the answer to this question turns out to be positive (under some natural assumptions, of course). Investigating the problem posed, we are mainly interested in the situation where F coincides with the space $C(T)$ of all real-valued continuous functions defined on T (we recall that T is a separable metric space). The special role of $C(T)$ is dictated by the following two circumstances:

1) it is important to have some characterizations (in terms of covariance functions) of continuous Gaussian processes which are mathematical models of numerous physical phenomena;

2) the space $C[0, 1]$ is universal in the class of all separable metric spaces, i.e. any separable metric space (in particular, any separable Banach space) can be realized as a subset of $C[0, 1]$.

In order to present a solution of the problem formulated above, we need the corresponding properties of the preordering $\leq_\xi$; more precisely, we need Lemma 4. First, let us discuss some rather simple situations.

Denote by the symbol $L_\infty(T)$ the family of all real-valued bounded functions on T. Since all processes from the class $\Gamma(T)$ are separable, the set

$$\{X \in L_\infty(T)\} = \{\omega \in \Omega \ : \ sup_{t \in T}|X_t(\omega)| < +\infty\}$$

is an event, i.e. belongs to the σ-algebra S of the basic probability space (Ω, S, P) on which the processes from $\Gamma(T)$ are given. So we may formulate the next result.

Theorem 5. *Let T be a separable metric space and let X and Y be any two processes from the class $\Gamma(T)$, such that*

$$Y \leq_\xi X.$$

Then the implication

$$P(\{X \in L_\infty(T)\}) = 1 \Rightarrow P(\{Y \in L_\infty(T)\}) = 1$$

holds true.

Proof. Suppose that $P(\{X \in L_\infty(T)\}) = 1$ and let us show that, in this case, the relations

$$sup_{t \in T} E(X_t^2) < +\infty, \quad sup_{t \in T} E(Y_t^2) < +\infty$$

are satisfied. Consider the first of them and suppose to the contrary that

$$sup_{t \in T} E(X_t^2) = +\infty.$$

Then, for each real number $r > 0$, we have

$$sup_{t \in T} P(\{|X_t| > r\}) = sup_{t \in T} (2/\pi)^{1/2} \int_{r_t}^{+\infty} exp(-u^2/2) du$$

where

$$r_t = \frac{r}{(E(X_t^2))^{1/2}}.$$

Our assumption implies that $inf\{r_t \ : \ t \in T\} = 0$, hence

$$sup_{t \in T} P(\{|X_t| > r\}) = 1.$$

Consequently, we get

$$P(\{X \notin L_\infty(T)\}) = lim_{r \to +\infty} P(\{sup_{t \in T}|X_t| > r\}) \geq$$

$$lim_{r \to +\infty}(sup_{t \in T} P(\{|X_t| > r\})) = 1,$$

which yields a contradiction.

In order to establish the second relation, fix a point $s \in T$ (assuming that $T \neq \emptyset$) and write

$$sup_{t\in T}E(Y_t^2) \leq 2sup_{t\in T}E((Y_t - Y_s)^2) + 2E(Y_s^2)$$
$$\leq 2sup_{t\in T}E((X_t - X_s)^2) + 2E(Y_s^2) \leq$$
$$8sup_{t\in T}E(X_t^2) + 2E(Y_s^2) < +\infty.$$

Now, it is clear that

$$\Delta(T) = 4sup_{t\in T}|E(X_t^2) - E(Y_t^2)| < +\infty.$$

Further, in virtue of Lemma 4, we have

$$P(\{Y \notin L_\infty(T)\}) = lim_{r\to+\infty}P(\{sup_{t\in T}|Y_t| > 2r\}) \leq$$
$$2lim_{r\to+\infty}P(\{sup_{t\in T}|X_t| > r\}) + 4lim_{r\to+\infty}exp(-r^2/\Delta(T))$$
$$= 2P(\{X \notin L_\infty(T)\}) = 0.$$

This completes the proof of Theorem 5.

The next statement can be established by using an analogous argument.

Theorem 6. *Let $X = (X_n)_{n\in\mathbf{N}}$ and $Y = (Y_n)_{n\in\mathbf{N}}$ be two random sequences from the class $\Gamma(\mathbf{N})$ and suppose that*

$$Y \leq_\xi X.$$

If, in addition, $lim_{n\to\infty}E(Y_n^2) = 0$, then

$$P(\{X \in c_0\}) = 1 \Rightarrow P(\{Y \in c_0\}) = 1.$$

Now, let us turn our attention to local behaviour of trajectories of stochastic processes.

Let T be an arbitrary metric space. For any real number $\delta > 0$ and for each $s \in T$, we denote by $U_\delta(s)$ the open δ-neighbourhood of s. Also, we denote by $C(T, s)$ the family of all those real-valued functions on T which are continuous at s. In other words, we may write

$$C(T, s) = \{(x_t)_{t\in T} \in \mathbf{R}^T \ : \ lim_{\delta\to 0}sup_{t\in U_\delta(s)}|x_t - x_s| = 0\}.$$

Now, if $X = (X_t)_{t\in T}$ is a separable stochastic process, then, for each $s \in T$, the set

$$\{X \in C(T, s)\} = \{\omega \in \Omega \ : \ X(\omega) \in C(T, s)\}$$

is an event, so we can speak of its probability. Note that

$$P(\{X \in C(T, s)\}) = 1 \Leftrightarrow$$

$$(\forall \varepsilon > 0)(lim_{\delta \to 0} P(\{sup_{t \in U_\delta(s)} |X_t - X_s| > \varepsilon\}) = 0).$$

The reader can check this equivalence without any difficulty.

We shall say that a given process X is continuous at $s \in T$ almost surely if

$$P(\{X \in C(T, s)\}) = 1.$$

Furthermore, we shall say that X is continuous almost surely if it is continuous at all points of T almost surely.

The following statement is true.

Theorem 7. *Let T be a separable metric space, let $X = (X_t)_{t \in T}$ and $Y = (Y_t)_{t \in T}$ be two processes from the class $\Gamma(T)$ and let*

$$Y \leq_\xi X.$$

Then, for every point $s \in T$, we have the implication

$$P(\{X \in C(T, s)\}) = 1 \Rightarrow P(\{Y \in C(T, s)\}) = 1.$$

Proof. Fix a point $s \in T$ and suppose that

$$P(\{X \in C(T, s)\}) = 1.$$

Then, according to the definition,

$$(1) \quad lim_{\delta \to 0} P(\{sup_{t \in U_\delta(s)} |X_t - X_s| > \varepsilon\}) = 0$$

for any $\varepsilon > 0$. Consider two auxiliary processes

$$X' = (X'_t)_{t \in T}, \qquad Y' = (Y'_t)_{t \in T}$$

defined by the formulae

$$X'_t = X_t - X_s \qquad (t \in T),$$

$$Y'_t = Y_t - Y_s \qquad (t \in T).$$

Obviously, we have

$$X'_u - X'_v = X_u - X_v \qquad (u \in T, \ v \in T),$$

$$Y'_u - Y'_v = Y_u - Y_v \qquad (u \in T, \ v \in T).$$

So, taking into account the relation $Y \leq_\xi X$, we may write

$$E((Y'_u - Y'_v)^2) = E((Y_u - Y_v)^2) \leq$$

$$E((X_u - X_v)^2) = E((X'_u - X'_v)^2)$$

for all $u \in T$ and $v \in T$. But this means that

$$Y' \leq_\xi X'.$$

Then, according to Lemma 4, for any $\varepsilon > 0$ and $\delta > 0$, we get

$$(2) \quad P(\{sup_{t \in U_\delta(s)} |Y'_t| > 2\varepsilon\}) \leq 2P(\{sup_{t \in U_\delta(s)} |X'_t| > \varepsilon\})$$

$$+ 4exp(-\varepsilon^2 / \Delta(\delta, s))$$

where

$$\Delta(\delta, s) = 4sup_{t \in U_\delta(s)} |E((X'_t)^2) - E((Y'_t)^2)|.$$

Furthermore, in view of $Y \leq_\xi X$, we obtain

$$\Delta(\delta, s) \leq 4sup_{t \in U_\delta(s)} E((X'_t)^2) + 4sup_{t \in U_\delta(s)} E((Y'_t)^2)$$

$$\leq 8sup_{t \in U_\delta(s)} E((X'_t)^2).$$

Also, denoting

$$\varepsilon_t = \frac{\varepsilon}{(E((X'_t)^2))^{1/2}} \qquad (t \in T)$$

and taking into account the equality

$$P(\{|X'_t| > \varepsilon\}) = (2/\pi)^{1/2} \int_{\varepsilon_t}^{+\infty} exp(-z^2/2) dz$$

and relation (1), we see that

$$lim_{\delta \to 0} sup_{t \in U_\delta(s)} E((X'_t)^2) = 0.$$

Consequently,

$$lim_{\delta \to 0} sup_{t \in U_\delta(s)} E((Y'_t)^2) = 0.$$

These two equalities imply

$$(3) \quad lim_{\delta \to 0} \Delta(\delta, s) = 0.$$

Now, it is evident that relations (1) - (3) lead to the equality

$$lim_{\delta \to 0} P(\{sup_{t \in U_\delta(s)} |Y_t - Y_s| > \varepsilon\}) = 0.$$

In other words,

$$P(\{Y \in C(T,s)\}) = 1,$$

which completes the proof of Theorem 7.

We have demonstrated that if a process X from the class $\Gamma(T)$ is continuous at some point $s \in T$ almost surely and majorizes (in the sense of $\leq_\xi$) another process Y from the same class, then Y turns out to be continuous at s almost surely. Nevertheless, we cannot deduce from this result that almost all trajectories of Y are continuous on T if almost all trajectories of X are continuous on T. This situation essentially differs from the corresponding situation for usual functions. Namely, for a process continuous almost surely, we cannot assert, in general, that almost all its trajectories are continuous. Indeed, since

$$P(\{Y \in C(T)\}) = P(\cap_{s \in T} \{Y \in C(T,s)\}),$$

it may happen, for an uncountable space T, that

$$(\forall s \in T)(P(\{Y \in C(T,s)\}) = 1)$$

and simultaneously

$$P(\{Y \in C(T)\}) = 0.$$

The corresponding examples are widely known and can be constructed without any difficulties.

Example 3. Suppose that our basic probability space coincides with the unit segment $[0,1]$ equipped with the standard Lebesgue measure, i.e.

$$(\Omega, S, P) = ([0,1], B([0,1]), \lambda).$$

Let us take $T = [0,1]$ endowed with the usual topology and define a stochastic process $Y = (Y_t)_{t \in T}$ by the formula

$$Y_t(\omega) = I_{[t,1]}(\omega) \qquad ((\omega, t) \in [0,1] \times [0,1]),$$

where $I_{[t,1]}$ denotes the indicator of $[t,1] \subset [0,1]$. It is not hard to check that, for any $\omega \in \Omega \setminus \{1\}$, the trajectory $(Y_t(\omega))_{t \in T}$ is a step-function with the unique discontinuity point $t = \omega$. Hence, Y possesses only one continuous trajectory. At the same time, for every point $s \in T$ and for arbitrary real numbers $\varepsilon > 0$ and $\delta > 0$, we have

$$P(\{\omega : sup_{|t-s|<\delta}|Y_t(\omega) - Y_s(\omega)| > \varepsilon\}) \leq 2\delta.$$

Consequently, if $\varepsilon > 0$ is fixed, then

$$lim_{\delta \to 0} P(\{sup_{|t-s|<\delta}|Y_t - Y_s| > \varepsilon\}) = 0,$$

which shows that our process Y is continuous almost surely.

It must also be noted that all trajectories of the process Y from Example 3 are left-hand continuous and have right-hand limits (at each point of $T = [0, 1]$). Therefore, Y is separable but no modification of Y possesses almost all continuous trajectories. In addition, Example 3 is not extraordinary because processes continuous almost surely, but with discontinuous trajectories, can be frequently met in applications and turn out to be natural models of various physical phenomena. For instance, a typical representative of such processes is the so-called Poisson process. In view of this circumstance, it is remarkable that, for an arbitrary Gaussian process with continuous covariance function, the continuity almost surely implies the continuity of almost all trajectories. This statement is a direct consequence of the well-known theorem due to Ito and Nisio, concerning the oscillation of a Gaussian process (see [97]). We are not going to present here the most general version of the above-mentioned oscillation theorem. We shall formulate and prove only one statement which is similar to Lemma 2 from Section 11.

First, we must introduce some auxiliary notions. Let T be a separable metric space and let $Z = (Z_t)_{t \in T}$ be a separable stochastic process given on a basic probability space (Ω, S, P). Let us define, for the process Z, the oscillation process

$$H(Z) = (H_t)_{t \in T}$$

by putting

$$H_t(\omega) = lim_{\varepsilon \to 0+} sup_{u,v \in U_\varepsilon(t)} |Z_u(\omega) - Z_v(\omega)| \qquad ((\omega, t) \in \Omega \times T).$$

Suppose that there exists a real-valued (possibly, extended) function $\alpha(Z) = (\alpha_t)_{t \in T}$ such that

$$P(\{\omega : H_t(\omega) = \alpha_t \ for \ all \ t \in T\}) = 1.$$

Then $\alpha(Z)$ is called the oscillation function of Z.

The next result is due to Ito and Nisio.

Theorem 8. *Let T be a σ-compact metric space and let $X = (X_t)_{t \in T}$ be a process from the class $\Gamma(T)$, whose covariance function is continuous. Then there exists an oscillation function $\alpha(X) = (\alpha_t)_{t \in T}$.*

Proof. We shall establish this result only for $T = [0, 1]$ (a similar argument works in the general case, so we leave some purely technical details to the reader).

Thus, we assume below that $T = [0, 1]$. Let us note that, since the covariance function of X is continuous, X can be represented in the form

$$(1) \quad X_t = \sum_{k \in \mathbf{N}} Z_k \phi_k(t) \qquad (t \in T),$$

where $(Z_k)_{k \in \mathbf{N}}$ is a sequence of independent $N(0,1)$-distributed Gaussian random variables and $(\phi_k)_{k \in \mathbf{N}}$ is a sequence of real-valued continuous functions on $[0, 1]$. The existence of such a representation was demonstrated in Section 10 (see Lemma 7 therein). Recall also that, for each $t \in [0, 1]$, the series in equality (1) converges P-almost everywhere and this equality holds true P-almost everywhere in Ω. Now, for any natural number n, define a stochastic process

$$X^{(n)} = (X_t^{(n)})_{t \in [0,1]}$$

by putting

$$X_t^{(n)} = X_t - \sum_{k \le n} Z_k \phi_k(t) \qquad (t \in [0, 1]).$$

It immediately follows from this definition that, for each $t \in [0, 1]$, we have

$$X_t^{(n)} = \sum_{k > n} Z_k \phi_k(t)$$

almost everywhere in Ω.

Now, let $Z' = (Z'_t)_{t \in [0,1]}$ be an arbitrary separable process. Fix a segment $[s, t] \subset [0, 1]$ and define

$$H_{s,t}(Z') = lim_{m \to +\infty} lim_{p \to +\infty} sup_{u,v \in\ ]s-1/m, t+1/m[,\ |u-v|<1/p} |Z'_u - Z'_v|.$$

Then $H_{s,t}(Z')$ is a random variable and is called the maximal oscillation of Z' on $[s, t]$. Indeed, in view of the separability of Z', this oscillation is stochastically equivalent to the random variable $H^*_{s,t}(Z')$ which is defined completely analogously but for which all u and v range over the set $Q \cap\ ]s - 1/m, t + 1/m[$ where Q is a fixed set of separability for Z' (e.g., Q can coincide with the set of all rational numbers of $[0, 1]$). Our goal is to demonstrate that $H_{s,t}(X)$ is degenerate for all s and t satisfying the relation $s \leq t$. First, let us observe that the continuity of all functions ϕ_k $(k \in \mathbf{N})$ implies

$$H^*_{s,t}(X) = H^*_{s,t}(X^{(n)}) \qquad (n \in \mathbf{N}).$$

Therefore, the random variable $H^*_{s,t}(X)$ is measurable with respect to the remainder σ-algebra associated with the sequence $(Z_k)_{k \in \mathbf{N}}$. Consequently, in virtue of the zero-one law, $H^*_{s,t}(X)$ is degenerate, i.e. there exists a constant

$$\alpha_{s,t} \in [0, +\infty]$$

such that

$$P(\{H_{s,t}(X) = \alpha_{s,t}\}) = P(\{H^*_{s,t}(X) = \alpha_{s,t}\}) = 1.$$

Hence

$$P(\{H_{s,t}(X) = \alpha_{s,t} \textit{ for all rational } s \textit{ and } t \ (s \leq t)\}) = 1.$$

Now, let us mention that, according to the definition, $H_{s,t}(X)$ is left-hand continuous at s and right-hand continuous at t. This yields at once that

$$P(\{H_{s,t}(X) = \alpha_{s,t} \textit{ for all } s \textit{ and } t \ (s \leq t)\}) = 1.$$

Finally, we put

$$\alpha(X) = (\alpha_t)_{t \in T} = (\alpha_{t,t})_{t \in T}.$$

Then a direct verification shows that $\alpha(X)$ is the oscillation function of X. Theorem 8 has thus been proved.

Let X be a process from the class $\Gamma(T)$, with continuous covariance function, and suppose that this process is continuous almost surely. Then it is evident that

$$\alpha_t = 0 \qquad (t \in T)$$

for $\alpha(X) = (\alpha_t)_{t \in T}$. Applying Theorem 8, we obtain that almost all trajectories of X are continuous. Therefore, the next statement can be formulated.

Theorem 9. *Let T be a σ-compact metric space, let X be a stochastic process from $\Gamma(T)$ and let the covariance function of X be continuous. Then the following two assertions are equivalent:*

1) almost all trajectories of X are continuous;

2) X is continuous almost surely.

In other words, we have

$$P(\{X \in C(T)\}) = 1 \Leftrightarrow (\forall s \in T)(P(\{X \in C(T,s)\}) = 1).$$

At last, we are able to present a comparison theorem for the continuity almost surely of processes from the class $\Gamma(T)$. This result is due to Marcus and Shepp (see [145]).

Theorem 10. *Let T be a σ-compact metric space, let X and Y be two processes from $\Gamma(T)$ and suppose that*

$$Y \leq_\xi X.$$

Then we have the implication

$$P(\{X \in C(T)\}) = 1 \Rightarrow P(\{Y \in C(T)\}) = 1.$$

Proof. Indeed, according to Theorem 9, the process X is continuous almost surely. Hence it is continuous in mean square. Since $Y \leq_\xi X$, we infer that Y is continuous in mean square, too. This implies that the covariance function of Y is continuous. Now, applying the relation $Y \leq_\xi X$ once more and taking into account Theorem 7, we see that Y is continuous almost surely, as well. It remains to utilize again Theorem 9.

Our brief survey of comparison methods (for Gaussian stochastic processes) is finished. In fact, we have demonstrated how various nontrivial theorems concerning the local behaviour of trajectories of such processes can be obtained by using the Anderson inequality and the Slepian inequality. Similarly to the material of the previous section, close relationships between the unimodality property of a centered Gaussian measure and the oscillation property of a centered Gaussian process played an essential role of our considerations. Simultaneous usage of these properties is, in our opinion, not only a methodological trick but also reflects some distinguishing features of Gaussian processes.

For more deep information about Gaussian processes and the regularity properties of their trajectories, see [75], [190] and [191].

EXERCISES

1. Show that $\leq_\zeta$ and $\leq_\eta$ are ordering relations, i.e. they are reflexive, transitive and antisymmetric.

2. Show that $\leq_\xi$ is a preordering relation, i.e. it is reflexive and transitive. Give also an example illustrating that, in general, $\leq_\xi$ is not antisymmetric.

3. Let K_1 and K_2 belong to $\Phi_+(T)$. Check that if the relations

$$K_2 \leq_\zeta K_1, \quad K_2 \leq_\eta K_1$$

are satisfied, then $K_1 = K_2$.

4. Let $(F, ||\cdot||)$ be a separable Banach space, let $(x_n)_{n\in\mathbf{N}}$ be a sequence of elements from F and let $(Z_n)_{n\in\mathbf{N}}$ be a sequence of independent centered Gaussian random variables. Fix any element

$$(a_n)_{n\in\mathbf{N}} \in l_\infty .$$

Prove that the implication

$$P(\{sup_{m\in\mathbf{N}}||\sum_{n\leq m} Z_n x_n|| < +\infty\}) = 1 \Rightarrow$$

$$P(\{sup_{m\in\mathbf{N}}||\sum_{n\leq m} a_n Z_n x_n|| < +\infty\}) = 1$$

holds true (apply Theorem 1 from this section).

5. Check that Theorem 2 fails to be true if, in its formulation, the ordering $\leq_\zeta$ is replaced by any of the preorderings $\leq_\eta$ and $\leq_\xi$. Specify the reasons of this circumstance.

6. Let $Y = (Y_n)_{n\in\mathbf{N}}$ be a centered Gaussian random sequence with the correlation coefficients $(r_{jn})_{j\in\mathbf{N},\ n\in\mathbf{N}}$ satisfying the inequality

$$lim_{m\to\infty}(sup_{j\geq m,\ n\geq m,\ j\neq n}\ r_{jn}) < 1.$$

Show that the relation

$$P(\{Y \in c_0\}) = 1$$

is true if and only if, for each $\varepsilon > 0$, the series

$$\sum_{n\in\mathbf{N}} exp(-\frac{\varepsilon}{E(Y_n^2)})$$

is convergent (apply the Slepian inequality).

7. Generalize Theorem 4 in the following manner: under its assumptions, for any real-valued positive function $f = (f_t)_{t \in T}$, the inequality

$$P(\{sup_{t \in T}(|Y_t| - f_t) > 0\}) \leq 2P(\{sup_{t \in T}(X_t + b_t Z_0 - f_t) > 0\})$$

is always fulfilled.

8. In connection with Theorem 6, give an example illustrating that the assumption $lim_{n \to \infty} E(Y_n^2) = 0$ in its formulation is essential for its validity.

Try to find a suitable analogue of this theorem in the situation when $lim_{n \to \infty} E(Y_n^2) = 0$ does not hold.

9. We recall that a process X from the class $\Gamma(\mathbf{R})$ is stationary if the values $K_X(t, s)$ $(t \in \mathbf{R},\ s \in \mathbf{R})$ of its covariance function depend only on the difference $s - t$.

Deduce from Theorem 8 the following alternative due to Belyaev: if a process $Y \in \Gamma(\mathbf{R})$ is stationary, then either trajectories of Y are continuous almost surely or they are unbounded almost surely on any neighbourhood of each point from $\mathbf{R}$.

13. Integration of vector-valued functions and optimal estimation of stochastic processes

In this section, we investigate the so-called prognosis problem which can be formulated as follows. Suppose that two stochastic processes X and Y are given, such that X is not accessible for a direct observation, but Y is completely accessible. Suppose also that there are some close relationships between these two processes. Then it is natural to estimate the values of X starting with the corresponding values of Y (in other words, having a certain information about the values of Y, we want to prognosticate the corresponding values of X). Problems of this type arise in various situations and play a significant role in mathematical statistics. In order to explain the deep connection of such problems with topics of our book, it is necessary to formulate these problems in a more precise manner and to specify the meaning of the phrase "prognostication".

Let (Ω, S, P) be a basic probability space and let X and Y be two random elements whose values belong to a topological vector space F. In other words, X and Y are measurable mappings acting from (Ω, S) into $(F, B(F))$ where $B(F)$ denotes, as usual, the Borel σ-algebra of F. In the sequel, some additional assumptions on F will be introduced, but now it suffices to suppose that F is a Polish topological vector space. We also assume that the process Y is completely described and, starting with its values, we try to estimate the corresponding values of X, i.e. we wish to find a random element $X' = X'(Y)$ (with values in the same space F) which approximates X in the best way.

If $F = \mathbf{R}$, then we deal with estimation of random variables.

If $F = \mathbf{R}^m$, where $m > 1$, then we deal with estimation of random vectors.

If $F = C[0,1]$, then we deal with estimation of stochastic processes having the continuous trajectories.

First of all, the class of random elements must be determined, in which we try to find the required estimate X'. For this purpose, it is reasonable to consider the σ-algebra S_Y generated by the given random element Y.

This σ-algebra is contained in S and its elements are sets of the form

$$\{Y \in C\} = \{\omega \in \Omega \ : \ Y(\omega) \in C\}$$

where C runs over $B(F)$. Denote by

$$L_0((\Omega, S_Y), F)$$

the family of all measurable mappings acting from (Ω, S_Y) into $(F, B(F))$. Clearly, S_Y involves all information which can be received by watching the random element Y. Therefore, it is sensible to take estimators for X from the family $L_0((\Omega, S_Y), F)$. The correctness of such an approach is confirmed by the following simple (but important) auxiliary statement.

Lemma 1. *Let F be a separable metrizable topological vector space. A random element Z in F belongs to $L_0((\Omega, S_Y), F)$ if and only if it admits a representation*

$$Z = g \circ Y = g(Y)$$

where g is some Borel mapping acting from F into itself.

Proof. If a random element Z is representable in the above-mentioned form, then the relation

$$(1) \quad Z \in L_0((\Omega, S_Y), F)$$

follows at once from the definition. In order to establish the converse assertion, suppose that (1) holds true for Z. First, let us consider the case when Z is simple, i.e.

$$Z = \sum_{n \in \mathbf{N}} z_n I_{A_n}$$

where $(z_n)_{n \in \mathbf{N}}$ is some sequence of elements from F and $\{A_n \ : \ n \in \mathbf{N}\}$ is some countable disjoint covering of Ω consisting of S_Y-measurable sets (here I_{A_n} denotes, as usual, the indicator of A_n). In virtue of the definition of S_Y, there exists a countable family $(C_n)_{n \in \mathbf{N}}$ of Borel sets in F, such that

$$A_n = \{Y \in C_n\} \qquad (n \in \mathbf{N}).$$

We may also assume that this family is disjoint. Define a Borel mapping

$$g \ : \ F \to F$$

by the formula

$$g(x) = \sum_{n \in \mathbf{N}} z_n I_{C_n}(x) \qquad (x \in F).$$

Then it is not hard to check that

$$Z = g \circ Y.$$

Consider now the general case of an arbitrary random element Z satisfying relation (1). Since F is separable, there exists a sequence $(Z_m)_{m \in \mathbf{N}}$ of simple random elements from $L_0((\Omega, S_Y), F)$, which is pointwise converging to Z, i.e. we have

$$lim_{m \to \infty} Z_m(\omega) = Z(\omega) \qquad (\omega \in \Omega).$$

Applying to each Z_m the result obtained above, we can find a Borel mapping

$$g_m \ : \ F \to F$$

such that

$$Z_m = g_m \circ Y = g_m(Y).$$

Let us observe that $lim_{m \to \infty} g_m(y)$ exists for any

$$y \in ran(Y) = Y(\Omega).$$

So we may define a mapping

$$g \ : \ F \to F$$

as follows: $g(x) = lim_{m \to \infty} g_m(x)$ if this limit exists for $x \in F$, and $g(x) = 0$ if this limit does not exist for $x \in F$. A direct verification shows that g is a Borel mapping and

$$Z = g \circ Y = g(Y),$$

which completes the proof of Lemma 1.

Thus, in our considerations below, the required estimators are assumed to be elements of the class $L_0((\Omega, S_Y), F)$. Further, in order to give a precise folmulation of the prognosis problem, we have to introduce a certain real-valued function on F which will indicate how near (to each other) are our process X and its estimator Z, i.e. this function evaluates the difference $X - Z$. Namely, suppose that a functional

$$\phi \ : \ F \to \mathbf{R}$$

is given and suppose that it is Borel. We shall say that ϕ is an error finctional (or a loss functional). This terminology reflects the circumstance that the random variable $\phi(X - Z)$ expresses some loss when a random element Z is taken instead of X. The mean error with respect to the given loss functional ϕ is

$$\Delta_\phi(Z) = E(\phi(X - Z))$$

and may be regarded as a criterion of estimation. Now, it is reasonable to introduce an optimal estimator as that Z for which the mean error is minimal. In other words, the following definition seems to be rather natural.

Definition 1. A random element

$$X' \in L_0((\Omega, S_Y), F)$$

is called an optimal estimator of a random element X (via watching a random element Y) with respect to a given loss functional ϕ if

$$\Delta_\phi(X') \leq \Delta_\phi(Z)$$

for all $Z \in L_0((\Omega, S_Y), F)$.

The following typical examples illustrate the notion just introduced.

Example 1. Let us put $F = \mathbf{R}$ and let

$$\phi(x) = x^2 \qquad (x \in \mathbf{R}),$$

$$\Delta_\phi(Z) = E((X - Z)^2).$$

Example 2. Let us put again $F = \mathbf{R}$ and let

$$\phi(x) = |x| \qquad (x \in \mathbf{R}),$$

$$\Delta_\phi(Z) = E(|X - Z|).$$

Example 3. Let again $F = \mathbf{R}$ and let $\delta > 0$ be fixed. Define a function ϕ from $\mathbf{R}$ into itself as follows: $\phi(x) = 0$ if $|x| \leq \delta$, and $\phi(x) = 1$ if $|x| > \delta$. Then we have

$$\Delta_\phi(Z) = P(\{|X - Z| > \delta\}).$$

Observe that the loss functions given in Examples 1 and 2 are particular cases of the loss function

$$\phi(x) = |x|^p \qquad (x \in \mathbf{R})$$

where p is a fixed strictly positive real number.

In all examples above $F = \mathbf{R}$. If F is an arbitrary topological vector space, then the role of loss functionals play, as a rule, compositions of functions ϕ from these examples with various seminorms on F.

Example 4. Let $F = C[0,1]$ and let ϕ be any function described in Examples 1 - 3. Define a functional

$$\psi : F \to \mathbf{R}$$

by putting

$$\psi(x) = \phi(sup_{t\in[0,1]}|x_t|) \qquad (x = (x_t)_{t\in[0,1]} \in C[0,1]).$$

Then it is not hard to see that ψ may be regarded as a loss functional on F.

Now, we wish to determine a class Φ of loss functionals, containing all the functionals considered above.

Definition 2. Let F be a topological vector space. We shall say that a functional

$$\phi : F \to [0,+\infty[$$

belongs to the class $\Phi = \Phi(F)$ if:

1) ϕ is a Borel mapping;

2) for each real number $r > 0$, the set

$$\{x \in F : \phi(x) < r\}$$

is convex and symmetric (with respect to the neutral element of F).

In general, the choice of a loss functional from the class Φ depends on specific features of the problem of optimal estimation for a given random element. If $F = \mathbf{R}$ and ϕ is as in Example 1, then the prognostication problem admits a complete solution. Namely, it turns out that, for a random variable X with finite variance, the optimal estimator of X (via watching a random variable Y) coincides with the conditional mean, i.e. conditional mathematical expectation, $E(X|Y)$. The precise definition of $E(X|Y)$ will be introduced below. Here we want to present a geometric interpretation of this fact.

Consider the space $L_2(\Omega, \mathbf{R})$ of all square integrable real-valued functions on Ω and denote by

$$L_2((\Omega, S_Y), \mathbf{R})$$

the subspace of $L_2(\Omega, \mathbf{R})$ consisting of all those random variables which are measurable with respect to S_Y. Clearly, these two spaces are Hilbert (with respect to the natural inner product inducing the mean square convergence). So we may examine here the estimation criterion

$$\Delta_\phi(Z) = E((X - Z)^2)$$

corresponding to the loss functional

$$\phi(Z) = ||Z||^2 \qquad (Z \in L_2(\Omega, \mathbf{R})).$$

In this situation, a random variable X (whose estimation we need) belongs to the space $L_2(\Omega, \mathbf{R})$ and we wish to find an optimal estimator in the subspace $L_2((\Omega, S_Y), \mathbf{R})$. Then the idea of orthogonality immediately leads us to the conclusion that the required optimal estimator coincides with the orthogonal projection of X on the subspace $L_2((\Omega, S_Y), \mathbf{R})$. But this orthogonal projection is exactly $E(X|Y)$.

For other loss functionals, the approach by means of the Hilbert space techniques is not applicable and optimal estimator may differ from the conditional mathematical expectation. The corresponding prognosis problem becomes much more difficult even in the case of a finite-dimensional F. Nevertheless, by using the unimodality property of centered Gaussian measures (i.e. the Anderson inequality for such measures), it is possible to prove that if X and Y are jointly Gaussian elements in a separable Banach space F and ϕ is an arbitrary functional from the class $\Phi(F)$, then an optimal estimator always coincides with $E(X|Y)$. In what follows we present a detailed proof of this nontrivial result. But first we need some preliminary material concerning means and conditional means of random elements in topological vector spaces. Actually, this material is due to the theory of integration of vector-valued functions.

Integration theory in finite-dimensional spaces can easily be reduced to the one-dimensional case. Indeed, if

$$X = (X_1, X_2, ..., X_m)$$

is a finite sequence of random variables (given on the same basic probability space), then we may put

$$E(X) = (E(X_1), E(X_2), ..., E(X_m)).$$

In this way, $E(X)$ becomes the mathematical expectation of the random vector X. It can easily be checked that the above-mentioned definition of $E(X)$ is correct, i.e. it does not depend on the choice of a coordinate system in $\mathbf{R}^m$ (in other words, $E(X)$ turns out to be invariant with respect to the group of all isometric transformations of $\mathbf{R}^m$).

Essential difficulties appear in those situations where we are forced to deal with infinite-dimensional topological vector spaces. There are different approaches to integration of vector-valued functions. One general approach was suggested by Pettis (see, e.g., [31], [56], [175] and [197]). We shall briefly discuss it.

As usual, we start with a basic probability space (Ω, S, P). In fact, we could take instead of (Ω, S, P) an arbitrary space equipped with a signed measure of finite variation (or a much more general measure). But we do not exploit such situations in the sequel, so we restrict ourselves to the case of (Ω, S, P). Let F be a topological vector space having the property that F^* separates the points of F. Let

$$X \ : \Omega \to F$$

be a weakly measurable mapping (this means that, for any functional h from F^*, the real-valued function $h(X) = h \circ X$ is S-measurable). We shall say, in accordance with the standard terminology, that X is scalarly integrable if

$$\int_{\Omega} |h(X(\omega))| P(d\omega) < +\infty$$

for all functionals $h \in F^*$. Equivalently, X is scalarly integrable if the real-valued function $h \circ X$ is Lebesgue integrable for any $h \in F^*$.

Suppose that X is scalarly integrable and that there exists a vector $y \in F$ satisfying the equality

$$h(y) = \int_{\Omega} h(X(\omega)) P(d\omega)$$

for each functional $h \in F^*$. Then X is called weakly integrable (or integrable in the Pettis sense) and y is called the Pettis integral of X. In this case, we usually write

$$y = \int_{\Omega} X(\omega) P(d\omega), \quad y = \int_{\Omega} X dP.$$

The uniqueness of y follows at once from our assumption that F^* separates the points of F. Hence only the question of the existence of Pettis integrals is nontrivial.

Returning to the terminology adopted in Section 6, we see that the class of weakly measurable mappings coincides with the class of weak random elements. Thus, it will be convenient to extend this terminology to Pettis integrals. Namely, if X is a weak random element intgrable in the Pettis sense, then its Pettis integral will be denoted by $E(X)$ and will be called the weak mean of X, or the weak mathematical expectation of X.

Let us point out some simple situations where the Pettis integral can easily be found by starting with its definition.

Example 5. Let X be a scalarly integrable random element in F. Suppose, in addition, that X is weakly symmetric, i.e. for any functional $h \in F^*$, the random variable $h \circ X$ is symmetric. Then, obviously,

$$E(h \circ X) = 0 \qquad (h \in F^*).$$

Consequently, the weak mean $E(X)$ exists and coincides with the neutral element of F.

Example 6. Let X be a weak centered Gaussian random element in F, i.e. for each functional $h \in F^*$, the random variable $h \circ X$ turns out to be Gaussian with $E(h \circ X) = 0$. Then, according to the preceding example, $E(X)$ exists and coincides with the neutral element of F.

Of course, these two examples are almost trivial. In the general situation, the question of the existence of a weak integral is rather complicated. The corresponding theory is discussed in detail in the monograph by Bourbaki [31] (cf. also [56], [175] and [197]). Here we need only some facts from this theory. First, let us indicate several simple properties of the Pettis integral.

I. If weak random elements X and Y are weakly integrable, then, for any real numbers a and b, the weak random element $aX + bY$ is weakly integrable, too, and we have the equality

$$E(aX + bY) = aE(X) + bE(Y).$$

II. Let F_1 be a topological vector space such that F_1^* separates the points of F_1 (note that F_1 may coincide with F), and let

$$U : F \to F_1$$

be a continuous linear mapping. Suppose also that X is a weakly integrable weak random element in F. Then the composition $U \circ X$ is a weakly integrable weak random element in F_1 for which we have

$$E(U \circ X) = U(E(X)).$$

III. Let X be a weak random element in F, let $\mu = \mu_X$ be the distribution of X (defined on the cylindrical σ-algebra of F) and suppose that there exists a weak integral $E(X)$. Then

$$h(E(X)) = \int_F h(x)\mu(dx)$$

for each functional $h \in F^*$.

Property III shows us that the question of the existence of Pettis integral $E(X)$ can be reduced to an analogous question for integral $\int_F x\mu(dx)$. Obviously, the latter integral is defined as a vector $y \in F$ satisfying the relation

$$h(y) = \int_F h(x)\mu(dx)$$

for any $h \in F^*$. In this way, we come to the geometric problem of the existence of a "centre of mass" for a probability measure μ given in the topological vector space F. Note that this problem admits a simple solution (under certain natural assumptions, of course). In particular, for Borel probability measures concentrated on compact subsets of F, the following statement is true (see [31] and [175]).

Theorem 1. *Let F be a topological vector space, let F^* separate the points of F and let C be a compact subset of F. Suppose, in addition, that the closed convex hull of C (i.e. the set $cl(conv(C))$) is also compact in F, and consider any Borel probability measure μ on C. Then there exists a unique weak integral*

$$y = \int_C x\mu(dx)$$

and $y \in cl(conv(C))$.

Proof. First, let us verify that the assertion of Theorem 1 holds in the case when F is a finite-dimensional topological vector space, i.e. when

$$F = \mathbf{R}^m$$

for some natural number m. Take any compact set $C \subset \mathbf{R}^m$ and suppose that μ is a Borel probability measure on C. Since C is bounded in $\mathbf{R}^m$, there exist integrals

$$y_j = \int_C x_j\mu(dx) \qquad (1 \leq j \leq m)$$

where, for each natural number $j \in [1, m]$, the symbol x_j denotes the j-th coordinate of a vector $x \in \mathbf{R}^m$. Now, it is clear that the vector

$$y = (y_1, y_2, ..., y_m) \in \mathbf{R}^m$$

will be the "centre of mass" (with respect to μ). It remains to demonstrate that

$$y \in cl(conv(C)).$$

Recall the classical theorem stating that the convex hull of any compact subset of $\mathbf{R}^m$ is compact, too (see Theorem 12 from Section 1). Hence we have to show that $y \in conv(C)$. Suppose otherwise, i.e. $y \notin conv(C)$. Then the point y and the compact convex set $conv(C)$ can be strongly separated by some affine hyperplane in $\mathbf{R}^m$. In other words, there exist real numbers $a_1, a_2, ..., a_m$ such that

$$\sum_{1 \leq j \leq m} a_j y_j < \sum_{1 \leq j \leq m} a_j x_j$$

for all vectors $(x_1, x_2, ..., x_m) \in conv(C)$. Integrating this inequality with respect to μ and taking into account the fact that μ is a probability measure, we come to the relation

$$\sum_{1 \leq j \leq m} a_j y_j < \sum_{1 \leq j \leq m} a_j y_j,$$

which yields a contradiction. Thus, y must be in $conv(C)$.

Now, let us consider the general case of a space F satisfying the assumptions of Theorem 1. In this case, take any finite family

$$H = (h_1, h_2, ..., h_m) \subset F^*$$

and denote by C_H the set of all those vectors $y \in cl(conv(C))$ for which the equalities

$$(1) \quad h_j(y) = \int_C h_j(x)\mu(dx) \qquad (1 \leq j \leq m)$$

are valid. Since all h_j $(j = 1, 2, ..., m)$ are continuous, the set C_H is a closed subset of $cl(conv(C))$ and hence is compact as well. Let us demonstrate that

$$C_H \neq \emptyset.$$

For this purpose, consider H as a mapping acting from F into $\mathbf{R}^m$, i.e. put

$$H(x) = (h_1(x), h_2(x), ..., h_m(x)) \qquad (x \in F).$$

Clearly, H is a continuous linear mapping. Consequently, the set $H(C)$ is compact in $\mathbf{R}^m$. Applying to this set the result proved above, we see that there exists a vector $z \in cl(conv(H(C)))$ satisfying the relation

$$z = \int_{H(C)} u\mu_H(du) = \int_C H(x)\mu(dx)$$

where μ_H denotes, as usual, the image of μ under H. Utilizing again the linearity and continuity of H, we may write

$$cl(conv(H(C))) = H(cl(conv(C))).$$

Hence there exists a vector $y \in cl(conv(C))$ such that

$$H(y) = z = \int_C H(x)\mu(dx).$$

In other words, (1) is fulfilled for y whence it follows that C_H is not empty.

Finally, one can easily observe that the family of nonempty compact sets

$$(C_H)_{H \subset F^*,\ H \text{ is finite}}$$

generates a filter, so we have

$$\cap_{H \subset F^*,\ H \text{ is finite}} C_H \neq \emptyset.$$

Then, for an element

$$y \in \cap_{H \subset F^*,\ H \text{ is finite}} C_H,$$

the relation

$$h(y) = \int_C h(x)\mu(dx) \qquad (h \in F^*)$$

holds true, which implies that y is the required weak integral.

The uniqueness of y trivially follows from our assumption that F^* separates the points of F.

Theorem 1 has thus been proved.

It is well known (see, e.g. Exercise 12 to Section 5) that the closed convex hull of a compact set in a topological vector space is not, in general, compact. However, if F is a Frechet space and C is an arbitrary compact subset of F, then $cl(conv(C))$ is compact, too. Consequently, the assertion of Theorem 1 holds for any Frechet space (in particular, for any Banach space). Also, it is evident that Theorem 1 essentially remains true for finite

Borel measures μ given on C. A weak integral exists in this case, too, but its value belongs to the image of $cl(conv(C))$ under the homothety of F whose coefficient is $\mu(C)$.

Let us recall the result of Ulam (see Section 6), according to which any σ-finite Borel measure given on a Polish topological space is concentrated on some σ-compact set and, moreover, is a Radon measure. By combining this result with Theorem 1, one can formulate various sufficient conditions for the existence of a centre of mass with respect to such measures. We are not going to discuss the conditions of this sort. Here we are interested in other methods for establishing the existence of weak integrals. Some sensible results may be obtained by starting with the existence theorems known from functional analysis. Indeed, let us remark, that, for a given scalarly integrable weak random element X in F, the mapping

$$h \to E(h \circ X) \qquad (h \in F^*)$$

is a linear functional on F^*. It was shown (see, e.g., [31]) that, for a wide class of topological vector spaces, this functional is also continuous (for example, this is true for all Frechet spaces F). Thus, the above-mentioned functional may be regarded as an element of the second conjugate space F^{**}. Now, if our original space F is reflexive, then we immediately obtain that a weak integral exists for any scalarly integrable weak random element in F (this result is due to Gelfand). Unfortunately, we do not have a similar result for more general spaces F. For such spaces, the problem of the existence of a weak integral turns out to be equivalent to the following classical problem from functional analysis: let

$$u \, : \, F^* \to \mathbf{R}$$

be a linear form; does there exist an element $y \in F$ such that $u(h) = h(y)$ for all $h \in F^*$? This problem was posed and investigated by Banach. It should be noted that if F is a separable Frechet space, then the solution of the problem can be formulated as follows: a vector y exists (and is unique) if and only if

$$lim_{n \to \infty} u(h_n) = 0$$

for each sequence $(h_n)_{n \in \mathbf{N}} \subset F^*$ satisfying the relation

$$lim_{n \to \infty} h_n(x) = 0 \qquad (x \in F),$$

i.e. for each sequence $(h_n)_{n \in \mathbf{N}} \subset F^*$ weakly converging to zero (the proof of this deep result can be found, e.g., in [29], [59] or [179]). Further, since in the problem of the existence of Pettis integral the corresponding linear form u is defined by

$$u(h) = E(h \circ X) \qquad (h \in F^*),$$

it suffices to establish that there exists a limit

$$lim_{n \to \infty} E(h_n \circ X)$$

for any sequence $(h_n)_{n \in \mathbf{N}} \subset F^*$ weakly converging to zero.

Suppose now that X is a weak Gaussian random element, and take an arbitrary sequence $(h_n)_{n \in \mathbf{N}} \subset F^*$ weakly converging to zero. Then the sequence $(h_n \circ X)_{n \in \mathbf{N}}$ of Gaussian random variables converges to zero with probability 1. Hence the sequence $(\nu_n)_{n \in \mathbf{N}}$ of Gaussian distributions of these random variables weakly converges to the Dirac probability measure ν on $\mathbf{R}$ (which is concentrated on the one-element set $\{0\}$). According to the Prokhorov theorem (see Section 7), the family of probability measures $(\nu_n)_{n \in \mathbf{N}}$ must be tight. But the last fact immediately implies the relations

$$sup_{n \in \mathbf{N}} |E(h_n \circ X)| < +\infty,$$

$$sup_{n \in \mathbf{N}} (E((h_n \circ X)^2) - (E(h_n \circ X))^2) < +\infty,$$

whence it follows that

$$lim_{r \to +\infty} \int_{|t|>r} |t| \nu_n(dt) = 0$$

uniformly with respect to $n \in \mathbf{N}$. So, taking into account once more the weak convergence of $(\nu_n)_{n \in \mathbf{N}}$ to ν, we may write

$$lim_{n \to \infty} u(h_n) = lim_{n \to \infty} E(h_n \circ X) = lim_{n \to \infty} \int_{\mathbf{R}} t \nu_n(dt) = 0.$$

In this manner, we have proved the next auxiliary statement.

Lemma 2. *Let X be a Gaussian random element in a separable Frechet space. Then there exists a weak mean $E(X)$.*

The considerations presented above show that, for the existence of Pettis integral, rather weak assumptions are sufficient. However, in some situations, the Pettis integral is not completely satisfactory. First of all, note that its definition is not constructive. Also, a number of important features which are valid for the standard Lebesgue integral fail to be true for the Pettis integral (see, e.g., Exercise 5 to this section). In many cases, it is more convenient to use the so-called Bochner integral (or the strong integral) whose properties are very similar to the properties of the Lebesgue integral. We shall introduce the Bochner integral for random elements in separable Banach spaces. The construction of this integral is rather analogous to the classical construction of the Lebesgue integral. First, we define the Bochner integral for simple random elements whose ranges are finite and then, by applying the corresponding limit operation, the integral will be extended on a much wider class of random elements (cf. [31], [56], [59]).

More precisely, let X be a simple random element in a separable Banach space F, i.e. X is representable in the form

$$X = \sum_{1 \le j \le m} x_j I_{A_j}$$

where $(x_j)_{1 \le j \le m}$ is some finite sequence of elements from our space and $\{A_j : 1 \le j \le m\}$ is a disjoint finite covering of Ω consisting of P-measurable sets (here the symbol I_{A_j} denotes, as usual, the indicator of A_j). Then the Bochner integral of X is defined as

$$\int_\Omega X(\omega)P(d\omega) = \sum_{1 \le j \le m} P(A_j)x_j .$$

Now, let

$$X : \Omega \rightarrow F$$

be an arbitrary random element in F. We shall say that X is strongly integrable (or integrable in the Bochner sense) if there exists a sequence $(X_n)_{n \in \mathbf{N}}$ of simple random elements in F, which converges P-almost everywhere to X and satisfies the relation

$$lim_{n \rightarrow \infty} \int_\Omega ||X(\omega) - X_n(\omega)||P(d\omega) = 0.$$

In such a case, it is not hard to show that there exists a limit of the sequence

$$(\int_\Omega X_n(\omega)P(d\omega))_{n \in \mathbf{N}}$$

and that this limit does not depend on the choice of a sequence $(X_n)_{n \in \mathbf{N}}$. It is called the strong integral of X (or the Bochner integral of X). The usual notation

$$\int_\Omega X(\omega)P(d\omega), \quad \int_\Omega X dP$$

is applicable to the strong integral as well (if, of course, no confusion arises with the same notation of the weak integral).

According to the well-known result (due to Lebesgue and Bochner), a random element X is strongly integrable if and only if

$$\int_{\Omega} ||X(\omega)||P(d\omega) < +\infty.$$

Moreover, for every strongly integrable random element X, we have the inequality

$$||\int_{\Omega} X(\omega)P(d\omega)|| \leq \int_{\Omega} ||X(\omega)||P(d\omega).$$

The reader can check these facts as a useful exercise.

Also, from the probabilistic viewpoint, it is convenient to say that the Bochner integral of X is the strong mean (or the strong mathematical expectation) of X. So we may utilize the notation $E(X)$, for the strong mean, in all those cases where no confusion is possible with the same notation of the weak mean. It can easily be verified that, for the strong mean $E(X)$, the relation

$$h(E(X)) = E(h \circ X)$$

holds true wnenever $h \in F^*$. Consequently, the strong mean of X (if it exists) coincides with the weak mean of X.

Note that many classical statements from the theory of Lebesgue integral remain valid for the Bochner integral (see, e.g., Exercise 6). In particular, if a random element X possesses the Bochner integral

$$\int_{\Omega} X(\omega)P(d\omega)$$

and C is an arbitrary event, then there exists a Bochner integral

$$\int_{C} X(\omega)P(d\omega)$$

and we have

$$\int_{C} X(\omega)P(d\omega) = \int_{\Omega} I_C(\omega)X(\omega)P(d\omega)$$

where I_C is the indicator of C. This natural property fails to be true for the Pettis integral (see Exercise 5).

We shall denote by $L_1(\Omega, F)$ the class of all those random elements in F which are integrable in the Bochner sense. Clearly, this class is a vector space under natural algebraic operations. Moreover, we may introduce a norm in this class by the formula

$$||X||_{L_1(\Omega,F)} = E(||X||) \qquad (X \in L_1(\Omega, F)).$$

Thus, we obtain the Banach space of all random elements integrable in the Bochner sense (as usual, P-equivalent random elements are identified). The family of all simple random elements is everywhere dense in this space.

We want to finish our brief survey of the integration theory of vector-valued functions with establishing the fact that any Gaussian random element in a separable Banach space is strongly integrable. In order to demonstrate this, we need one auxiliary proposition concerning the integrability of the norm of a Gaussian random element. We shall prove this proposition by using the method due to Fernique (see [75]).

Lemma 3. *Let $(F, ||\cdot||)$ be a separable Banach space and let X be a centered Gaussian random element in F. Then, for every real number $p > 0$, we have*

$$E(||X||^p) < +\infty.$$

Proof. It was pointed out in Exercise 9 from Section 10 that if X_1 and X_2 are any two independent copies of X, then the random elements

$$Y_1 = (1/2)^{1/2}(X_1 + X_2), \quad Y_2 = (1/2)^{1/2}(X_1 - X_2)$$

are independent copies of X, too. Consequently, for all real numbers s and t, we may write

$$P(\{||X|| \le s\})P(\{||X|| > t\}) = P(\{||Y_1|| \le s, \ ||Y_2|| > t\}).$$

Further, since

$$2^{1/2}||X_i|| \ge ||Y_2|| - ||Y_1|| \qquad (i = 1, 2),$$

we have the inclusion

$$\{||Y_1|| \le s, \ ||Y_2|| > t\} \subset \{||X_1|| > (1/2)^{1/2}(t-s), \ ||X_2|| > (1/2)^{1/2}(t-s)\}.$$

Applying the independence of X_1 and X_2 and taking into account the fact that these two random elements are equidistributed with X, we easily infer

$$P(\{||X|| \le s\})P(\{||X|| > t\}) \le (P(\{||X|| > (t-s)/2\}))^2.$$

Now, there always exists a real number $s_0 > 0$ such that

$$q = 2P(\{||X|| > s_0\}) < 1.$$

Hence, for any $t > 0$, the inequality

$$P(\{||X|| > 2t + s_0\}) \leq 2(P(\{||X|| > t\}))^2$$

holds true. The subsequent iteration of this inequality yields

$$P(\{||X|| > 2^{m+1}s_0\}) \leq P(\{||X|| > 2^{m+1}s_0 - s_0\}) \leq$$

$$2(P(\{||X|| > 2^m s_0 - s_0\}))^2 \leq \ \dots \ \leq 2^1 \cdot 2^2 \cdot \dots \cdot 2^{2^{m-1}} \cdot$$

$$(P(\{||X|| > s_0\}))^{2^m} \leq q^{2^m}.$$

The last relation implies that

$$E(||X||^p) = \int_0^{+\infty} t^p dP(\{||X|| \leq t\}) \leq (4s_0)^p +$$

$$\sum_{m\geq 1} (2^{m+2}s_0)^p P(\{||X|| > 2^{m+1}s_0\}) \leq$$

$$(4s_0)^p + \sum_{m\geq 1} (2^{m+2}s_0)^p q^{2^m} < +\infty$$

for each $p > 0$. This ends the proof of the lemma.

Let us observe that the method applied in the preceding argument enables to obtain a much stronger result (cf. Exercise 7 to this section). However, for our further purposes, the assertion of Lemma 3 is completely sufficient.

Now, we are able to formulate and prove the next statement.

Theorem 2. *Let F be a separable Banach space and let X be a Gaussian random element in F. Then X is integrable in the Bochner sense, i.e.*

$$X \in L_1(\Omega, F).$$

Proof. In view of Lemma 2, our random element X possesses the weak mean $E(X)$. Therefore, X can be represented in the form

$$X = E(X) + (X - E(X)).$$

For each functional $h \in F^*$, we have

$$h(X - E(X)) = h \circ X - E(h \circ X)$$

whence it follows that all random variables

$$h(X - E(X)) \qquad (h \in F^*)$$

are centered Gaussian. This shows us that the random element

$$Y = X - E(X)$$

is centered Gaussian, too. Hence, according to Lemma 3,

$$E(||Y||) < +\infty.$$

Then, obviously,

$$E(||X||) \leq E(||Y||) + ||E(X)|| < +\infty,$$

which implies that X is integrable in the Bochner sense. Theorem 2 has thus been proved.

Our further discussion is devoted to some elementary properties of conditional means. They will be defined only for random elements integrable in the strong sense. Let us recall that the concept of a conditional mean was first introduced by Kolmogorov in his classical monograph [121] (see also [58]).

Consider the space $L_1(\Omega, F)$ where F is a separable Banach space. Let S_0 be some σ-subalgebra of S. The symbol

$$L_1((\Omega, S_0), F)$$

will denote the class of all S_0-measurable strongly integrable random elements in F. Evidently, $L_1((\Omega, S_0), F)$ is a closed vector subspace of $L_1(\Omega, F)$.

Let $X \in L_1(\Omega, F)$. We shall say that a random element Z from the space $L_1((\Omega, S_0), F)$ is a conditional mean of X (or conditional mathematical expectation of X) with respect to the σ-algebra S_0 if, for each event $C \in S_0$, we have

$$\int_C X dP = \int_C Z dP.$$

In this case, we denote

$$Z = E(X|S_0).$$

Let us stress that if Z_1 and Z_2 are any two conditional means of X (with respect to S_0), then, for every functional $h \in F^*$, we have

$$\int_C (h \circ Z_1) dP = \int_C (h \circ Z_2) dP \qquad (C \in S_0)$$

whence it follows that

$$h(Z_1 - Z_2) = 0 \qquad (h \in F^*)$$

almost surely. Taking into account the fact that F^* contains a countable subset separating the points of F, we easily come to the relation

$$P(\{Z_1 = Z_2\}) = 1$$

which shows us that the random elements Z_1 and Z_2 are stochastically equivalent. Thus, we can conclude that the conditional mean is unique (up to the P-equivalence). The main problem is to establish the existence of a conditional mean. The next classical result gives a positive solution to this problem (see [121]).

Theorem 3. *For each random element $X \in L_1(\Omega, F)$ and for any σ-algebra $S_0 \subset S$, there exists $E(X|S_0)$.*

Proof. First, let us demonstrate the theorem for any random variable $X \in L_2(\Omega, \mathbf{R})$ where $L_2(\Omega, \mathbf{R})$ denotes the Hilbert space of all square integrable random variables. Accordingly, the symbol $L_2((\Omega, S_0), \mathbf{R})$ will denote the Hilbert subspace of this space, consisting of S_0-measurable random variables. Let

$$Pr_0 \; : \; L_2(\Omega, \mathbf{R}) \rightarrow L_2((\Omega, S_0), \mathbf{R})$$

be the canonical orthogonal projection and let

$$Z = Pr_0(X).$$

Clearly, we have

$$Z \in L_2((\Omega, S_0), \mathbf{R}) \subset L_1((\Omega, S_0), \mathbf{R}).$$

In addition, for any event $C \in S_0$, the indicator I_C belongs to $L_2((\Omega, S_0), \mathbf{R})$ and, in terms of the inner product, we may write

$$\int_C X dP = < X, I_C > = < X, Pr_0(I_C) > = < Pr_0(X), I_C > =$$

$$< Z, I_C > = \int_C Z dP.$$

Thus, the conditional mean of X exists and is equal to $Pr_0(X)$. In particular, there exist conditional means for all indicators of events from S. Besides, if $X \geq 0$ almost surely, then $E(X|S_0) \geq 0$ almost surely. Consequently, if A is an arbitrary event from S, then

$$E(|E(I_A|S_0)|) = E(E(I_A|S_0)) = E(I_A) = P(A)$$

whence it follows that

$$(1) \quad E(|E(I_A|S_0)|) = P(A) \qquad (A \in S).$$

Now, let X be a simple random element in F. Then X can be represented in the form

$$X = \sum_{1 \leq j \leq m} I_{A_j} x_j$$

where $(x_j)_{1 \leq j \leq m}$ is a finite subfamily of F and $\{A_j : 1 \leq j \leq m\}$ is a finite disjoint covering of Ω consisting of P-measurable sets. Then, taking into account the linearity of the Bochner integral and applying the preceding result, we easily obtain the existence of $E(X|S_0)$ and the corresponding formula

$$E(X|S_0) = \sum_{1 \leq j \leq m} E(I_{A_j}|S_0) x_j.$$

Moreover, in virtue of relation (1), we get the inequality

$$(2) \quad E(||E(X|S_0)||) \leq E(||X||).$$

Consider now the general situation when $X \in L_1(\Omega, F)$. In this case, there exists a sequence $(X_n)_{n \in \mathbf{N}}$ of simple random elements in F, such that

$$lim_{n \to \infty} E(||X - X_n||) = 0.$$

As stated above, for each X_n, we have the conditional mean

$$Z_n = E(X_n|S_0),$$

i.e. for all natural numbers n, the relations

$$(3) \quad \int_C X_n dP = \int_C Z_n dP \qquad (C \in S_0)$$

are fulfilled. Further, if n and k are arbitrary natural numbers, then the difference $X_n - X_k$ is a simple random element in F and, according to (2), the inequality

$$E(||Z_n - Z_k||) \leq E(||X_n - X_k||)$$

holds true. This implies that $(Z_n)_{n\in\mathbf{N}}$ is a Cauchy sequence in the Banach space $L_1(\Omega, F)$ which converges (in the metric of this space) to some random element $Z \in L_1(\Omega, F)$. But, since all Z_n $(n \in \mathbf{N})$ belong to $L_1((\Omega, S_0), F)$, we conclude that Z belongs to $L_1((\Omega, S_0), F)$, too. Tending an index n in (3) to infinity and applying a suitable analogue of the Lebesgue theorem on majorized convergence, we obtain

$$\int_C XdP = \int_C ZdP \qquad (C \in S_0),$$

which shows us that Z is the conditional mean of X with respect to S_0. Theorem 3 has thus been proved.

The argument presented above does not rely on the existence of conditional means for random variables. Therefore, our proof involves the case $F = \mathbf{R}$. However, in this case the proof can be done directly by using the classical Radon-Nikodym theorem (see, e.g., [155] or [180]).

Note also that if S_0 coincides with S, then

$$E(X|S_0) = X$$

for all random elements $X \in L_1(\Omega, F)$. If S_0 is the trivial σ-algebra (i.e. $S_0 = \{\emptyset, \Omega\}$), then

$$E(X|S_0) = E(X)$$

for all $X \in L_1(\Omega, F)$.

Let us observe some other basic properties of conditional means (with respect to a fixed σ-algebra $S_0 \subset S$).

1. If X_1 and X_2 are arbitrary strongly integrable random elements, and a_1 and a_2 are any real numbers, then

$$E((a_1X_1 + a_2X_2)|S_0) = a_1E(X_1|S_0) + a_2E(X_2|S_0).$$

2. If X is an arbitrary strongly integrable random element, then

$$E(E(X|S_0)) = E(X).$$

3. If S_1 is a σ-subalgebra of S such that $S_0 \subset S_1$, then, for every strongly integrable random element X, we have

$$E(E(X|S_1)|S_0) = E(X|S_0).$$

4. If a random element $X \in L_1(\Omega, F)$ is independent with respect to the σ-algebra S_0, then

$$E(X|S_0) = E(X).$$

5. If $X \in L_1(\Omega, F)$ and g is an S_0-measurable random variable such that $gX \in L_1(\Omega, F)$, then

$$E(gX|S_0) = gE(X|S_0).$$

6. Let $X \in L_1(\Omega, F)$ and let $h \in F^*$. Then there exists $E((h \circ X)|S_0)$ and

$$h(E(X|S_0)) = E((h \circ X)|S_0).$$

7. For all $X \in L_1(\Omega, F)$, we have

$$||E(X|S_0)|| \leq E(||X|||S_0).$$

These properties easily follow from the definition, so we leave their checking to the reader. We wish to emphasize once more that the conditional means are defined up to the canonical P-equivalence relation between random elements. Hence assertions 1 - 7 above are valid only with P-probability 1.

The next statement seems to be much more interesting.

Lemma 4. *Let $(S_n)_{n \in \mathbf{N}}$ be an increasing (with respect to the inclusion relation) sequence of σ-subalgebras of S and let S_∞ denote the σ-algebra generated by the algebra $\cup_{n \in \mathbf{N}} S_n$. Then, for every random variable X from $L_1(\Omega, \mathbf{R})$, the following equality is satisfied:*

$$lim_{n \to \infty} E(|E(X|S_\infty) - E(X|S_n)|) = 0.$$

Proof. In view of property 3 of conditional means, for each natural number n, we have

$$E(X|S_n) = E(E(X|S_\infty)|S_n).$$

Therefore, we may assume without loss of generality that the σ-algebra S_∞ coincides with S. Then we must show that

$$(1) \quad lim_{n \to \infty} E(|X - E(X|S_n)|) = 0.$$

For this purpose, let us consider the sequence of spaces

$$(L_n)_{n \in \mathbf{N}} = (L_1((\Omega, S_n), \mathbf{R}))_{n \in \mathbf{N}}.$$

Evidently,

$$L_n \subset L_{n+1} \qquad (n \in \mathbf{N})$$

and the vector space

$$L_\infty = \cup_{n \in \mathbf{N}} L_n$$

is everywhere dense in $L_1(\Omega, \mathbf{R})$. Hence, for each $\varepsilon > 0$, there exists a random variable $X_\varepsilon \in L_\infty$ satisfying the inequality

$$E(|X - X_\varepsilon|) < \varepsilon.$$

Since $X_\varepsilon \in L_n$ for all those natural numbers n which are greater than some fixed number $n(\varepsilon)$, we can write

$$E(|E(X|S_n) - X|) \leq E(|E(X|S_n) - X_\varepsilon|) + E(|X - X_\varepsilon|) =$$

$$E(|E((X - X_\varepsilon)|S_n)|) + E(|X - X_\varepsilon|) \leq 2E(|X - X_\varepsilon|) < 2\varepsilon.$$

This immediately leads to the validity of relation (1), and the lemma has thus been proved.

Returning to the problem of optimal estimation of a random process, let us introduce the conditional mean $E(X|Y)$ of an F-valued random element X, with respect to a random element Y, by the formula

$$E(X|Y) = E(X|S_Y)$$

where S_Y denotes the σ-subalgebra of S generated by Y (in this definition, we do not assume that Y is a random element in the same space F). Further, if X and Y are random elements in F, then $E(X|Y)$ is a random element measurable with respect to S_Y, and we may apply Lemma 1 according to which there exists a Borel mapping

$$g \; : \; F \to F$$

such that

$$E(X|Y) = g \circ Y = g(Y).$$

Note that g is determined uniquely up to a μ_Y-measure zero set, where μ_Y denotes the distribution of Y. Actually, the mapping g is utilized for prognostication of X via watching Y. In statistics, g is usually called the regression of X with respect to Y and the notation

$$g(y) = E(X|Y = y) \qquad (y \in F)$$

is commonly used.

A careful analysis of the proof of Theorem 3 shows us that if X and Y are any two square integrable random variables, then an optimal estimator of X (via watching Y) coincides with the conditional mean $E(X|Y)$. Indeed, it was demonstrated there that $E(X|Y)$ is the orthogonal projection of X on the subspace $L_2((\Omega, S_Y), \mathbf{R})$. Consequently, we may write

$$E((X - E(X|Y))^2) = ||X - E(X|Y)||^2 \leq$$

$$||X - Z||^2 = E((X - Z)^2)$$

for all random elements $Z \in L_2((\Omega, S_Y), \mathbf{R})$.

An analogous result can be obtained in the case when the criterion of estimation is given by the formula

$$Z \to min_{h \in F^*} E((h(X - Z))^2),$$

because we have the inequality

$$E((h(X - E(X|Y)))^2) \leq E((h(X - Z))^2)$$

for each functional $h \in F^*$ and for any random element Z satisfying the relation

$$h \circ Z \in L_2((\Omega, S_Y), \mathbf{R}).$$

However, we are interested here in obtaining more strong criteria of estimation, which will be suitable for all error functionals from the class Φ introduced earlier. Some additional assumptions on random elements X and Y are necessary in this case (motivated by probabilistic features of the problems discussed). For us, one of appropriate assumptions is the assumption that X and Y are jointly Gaussian.

In this connection, let us recall that F-valued random elements X and Y are jointly Gaussian if, for any functionals h and g from F^*, the random variables $h(X)$ and $g(Y)$ have joint Gaussian distribution, i.e. the random vector $(h(X), g(Y))$ is Gaussian. This condition is equivalent to the following: for every finite family

$$(h_1, h_2, ..., h_n, g_1, g_2, ..., g_m)$$

of functionals from F^*, the random vectors

$$(h_1(X), h_2(X), ..., h_n(X)), \quad (g_1(Y), g_2(Y), ..., g_m(Y))$$

are jointly Gaussian.

For given jointly Gaussian random elements X and Y, a certain representation of X, by means of the conditional expectation $E(X|Y)$, will be useful in the sequel.

Lemma 5. *Let X and Y be two jointly Gaussian random elements in a separable Banach space F. Then*

$$X = E(X|Y) + Z$$

where Z is a centered Gaussian element in F independent with respect to Y.

Proof. We begin with the case of finite-dimensional random elements X and Y. Let

$$X = (X_1, X_2, ..., X_n), \quad Y = (Y_1, Y_2, ..., Y_m)$$

be any two jointly Gaussian random vectors in $\mathbf{R}^n$ and $\mathbf{R}^m$ respectively (here n may differ from m). Hence the random vector

$$(X_1, X_2, ..., X_n, Y_1, Y_2, ..., Y_m)$$

is Gaussian in the space $\mathbf{R}^{n+m}$. Let us put

$$e_1 = E(X), \quad e_2 = E(Y),$$

$$K^{(1)} = cov(X), \quad K^{(2)} = cov(Y),$$
$$K^{(1,2)} = cov(X, Y) = E((X - e_1)^*(Y - e_2)),$$

where the symbol $(z_1, z_2, ..., z_n)^*$ denotes the column-vector transpose of the row-vector $(z_1, z_2, ..., z_n)$. Obviously, the means e_1 and e_2 are some vectors in $\mathbf{R}^n$ and $\mathbf{R}^m$ respectively, $K^{(1)}$ and $K^{(2)}$ are the covariance matrices of X and Y, respectively, of sizes $n \times n$ and $m \times m$, and $K^{(1,2)}$ is the mutual covariance matrix of X and Y whose size is $n \times m$ and whose terms are

$$K_{ij}^{(1,2)} = E((X_i - E(X_i))(Y_j - E(Y_j)))$$

for all natural numbers $i \in [1, n]$ and $j \in [1, m]$. We may assume without loss of generality that the covariance matrix $K^{(2)}$ is nondegenerate (otherwise, Y can be replaced by the random vector of less dimension). Thus, we suppose in our further considerations that there exists the inverse matrix of $K^{(2)}$. Now, let us define

$$(1) \quad K = K^{(1,2)}(K^{(2)})^{-1}, \quad X' = e_1 + K(Y - e_2).$$

Our goal is to demonstrate that X' coincides with $E(X|Y)$. For this purpose, let us observe that the pair of random vectors

$$(X - X', Y - e_2)$$

is obtained as the result of an affine transformation of the jointly Gaussian pair (X, Y). Consequently, $X - X'$ and $Y - e_2$ are jointly Gaussian vectors, too. A direct verification shows that

$$E((X - X')^*(Y - e_2)) = 0,$$

i.e. the mutual covariance matrix of these random vectors is zero. Hence any component of $X - X'$ is noncorrelated with any component of $Y - e_2$, which implies that these random vectors are independent. Therefore, X can be represented in the form

$$X = X' + (X - X')$$

where $X - X'$ is independent with respect to $Y - e_2$ (equivalently, with respect to Y). Simultaneously, the random vector X' being the image of Y, under an affine transformation, is S_Y-measurable. So, applying the corresponding properties of conditional means, we get

$$E(X|Y) = E(X'|Y) + E((X - X')|Y) = X' + E(X - X') =$$

$$X' - E(K(Y - e_2)) = X' - K(E(Y - e_2)) = X'.$$

Thus, we have established that

$$(2) \quad X' = E(X|Y)$$

and that X is representable in the form

$$X = E(X|Y) + Z,$$

where Z and Y are independent. At the same time, we obtain that

$$E(X|Y), \quad X - E(X|Y)$$

are independent Gaussian random vectors. It remains to mention the equalities

$$E(Z) = E(X) - E(E(X|Y)) = E(X) - E(X) = 0$$

which show that Z is centered.

Let us also point out that relations (1) and (2) enable to express visually the conditional mean $E(X|Y)$ for jointly Gaussian finite-dimensional random vectors X and Y (this fact is frequently helpful in various practical problems).

Now, we are ready to prove the statement of Lemma 5 in the general case. For any two jointly Gaussian elements X and Y given in a separable Banach space F, consider the random vector

$$Z = X - E(X|Y).$$

This definition of Z is correct since, according to Theorems 2 and 3, the conditional mean $E(X|Y)$ does exist. Evidently,

$$X = E(X|Y) + Z.$$

Choose a countable subset $H = (h_n)_{n \in \mathbf{N}}$ of F^* separating the points of F. Then, as we know,

$$(3) \quad B(F) = B(F, H).$$

For each natural number n, denote by S_n the σ-subalgebra of S generated by the family of random variables

$$(h_0(Y), h_1(Y), ..., h_n(Y)).$$

Clearly, the sequence $(S_n)_{n \in \mathbf{N}}$ is increasing by inclusion and, taking into account relation (3), we have

$$S_Y = \sigma(\cup_{n \in \mathbf{N}} S_n).$$

It follows from this, in virtue of Lemma 4, that the sequence of conditional means

$$(4) \quad (E(h(X)|(h_0(Y), h_1(Y), ..., h_n(Y))))_{n \in \mathbf{N}}$$

converges (in the metric of $L_1(\Omega, \mathbf{R})$) to $E(h(X)|Y)$ for any $h \in F^*$. According to our assumption, X and Y are jointly Gaussian whence it follows that the random variable $h(X)$ and the random vector

$$(h_0(Y), h_1(Y), ..., h_n(Y))$$

are jointly Gaussian, too. Hence, as proved above, the conditional mean

$$E(h(X)|(h_0(Y), h_1(Y), ..., h_n(Y)))$$

is the image of $(h_0(Y), h_1(Y), ..., h_n(Y))$ under an affine mapping and also has Gaussian distribution. Moreover, the sequence in (4) is a family of jointly Gaussian random variables, so its limit $E(h(X)|Y)$ is a Gaussian random variable. Applying one of the properties of conditional means, we may write

$$h(E(X|Y)) = E(h(X)|Y) \qquad (h \in F^*).$$

This implies that $E(X|Y)$ is a Gaussian random element in F. The same argument leads to a more general result. Namely, it turns out that the random elements

$$X, \ Y, \ E(X|Y)$$

have joint Gaussian distribution. Consequently, $X - E(X|Y)$ is a centered Gaussian element, and the random elements $X - E(X|Y)$ and Y are jointly Gaussian. It remains to demonstrate that $X - E(X|Y)$ and Y are independent as well. Fix any two functionals g_1 and g_2 from F^*. Then, applying again the corresponding properties of conditional means, we get the equalities

$$E(g_1(X - E(X|Y))g_2(Y)) =$$
$$E((g_1(X) - E(g_1(X)|Y))g_2(Y)) =$$
$$E(g_1(X)g_2(Y)) - E(g_2(Y)E(g_1(X)|Y)) =$$
$$E(g_1(X)g_2(Y)) - E(E(g_1(X)g_2(Y)|Y)) = 0.$$

Thus, we see that the random variables $g_1(X - E(X|Y))$ and $g_2(Y)$ are noncorrelated for all functionals g_1 and g_2 from F^*. This yields that the random elements $X - E(X|Y)$ and Y are independent, and the proof of the lemma is completed.

By the way, the preceding argument establishes also that $E(X|Y)$ is a Gaussian random element, and some constructive procedure for calculating $E(X|Y)$ is pointed out.

Our consideration of the question concerning optimal estimation of the first component of a jointly Gaussian pair (X, Y) (by watching the second component) is almost finished. In order to obtain a complete solution of this question, we appeal again to the Anderson inequality for centered Gaussian measures given in separable Banach spaces. Namely, we can formulate and prove the following result (see [36]).

Theorem 4. *Let X and Y be two jointly Gaussian random elements in a separable Banach space F. Then, for every loss functional $\phi \in \Phi = \Phi(F)$, the conditional mean $E(X|Y)$ turns out to be an optimal estimator of X (by watching Y).*

Proof. Fix any $\phi \in \Phi$. Actually, we must show that

$$(1) \quad E(\phi(X - E(X|Y))) \leq E(\phi(X - Z))$$

for each $Z \in L_0((\Omega, S_Y), F)$. Note that, since $\phi(x) \geq 0$ for all $x \in F$, the mathematical expectations in (1) do exist (however, they may be equal to $+\infty$). Let Z be an arbitrary S_Y-measurable random element in F. It will be convenient to represent $X - Z$ in the following form:

$$X - Z = Z_1 + Z_2,$$

where Z_1 and Z_2 are defined by the equalities

$$Z_1 = X - E(X|Y), \quad Z_2 = E(X|Y) - Z.$$

The random element Z_2 (being the difference of two S_Y-measurable random elements) is S_Y-measurable. Consequently, in view of Lemma 5, the random elements Z_1 and Z_2 are independent and, in addition to this, Z_1 is centered. Relation (1) is obviously equivalent to the next relation:

$$(2) \quad E(\phi(Z_1)) \leq E(\phi(Z_1 + Z_2)).$$

We shall prove (2) utilizing the Anderson inequality for a centered Gaussian measure (see Theorem 3 from Section 10).

Let us denote by μ_1 and μ_2 the distributions (in F) of Z_1 and Z_2 respectively. Then μ_1 and μ_2 are Gaussian measures on F and, besides, μ_1 is centered. Let us check that

$$(3) \quad \int_F \phi(x)\mu_1(dx) \leq \int_F \phi(x + y)\mu_1(dx)$$

for each vector $y \in F$. Suppose first that ϕ is bounded, i.e. there exists a real number $a \in]0, +\infty[$ such that

$$\phi(x) \leq a \qquad (x \in F).$$

Then we have the equality

$$\int_F \phi(x)\mu_1(dx) = lim_{n\to\infty} \sum_{0\leq k\leq m(n)} (1/n)\mu_1(C(k,n))$$

where $m(n)$ denotes the smallest natural number among all those natural numbers which are greater than or equal to na, and

$$C(k,n) = \{x \in F \ : \ \phi(x) \geq k/n\} \qquad (n \in \mathbf{N} \setminus \{0\}, \ k = 0, 1, ..., m(n)).$$

According to Definition 2, all sets $F \setminus C(k,n)$ are convex and symmetric. Hence, applying the Anderson inequality, we obtain

$$\int_F \phi(x)\mu_1(dx) = lim_{n\to\infty} \sum_{0\leq k\leq m(n)} (1/n)\mu_1(C(k,n)) \leq$$

$$lim_{n \to \infty} \sum_{0 \le k \le m(n)} (1/n)\mu_1(C(k,n) - y) = \int_F \phi(x+y)\mu_1(dx).$$

Suppose now that our functional ϕ is unbounded. In this case, for any $a > 0$, we introduce a new functional

$$\phi_a \; : \; F \to \mathbf{R}$$

by putting $\phi_a(x) = \phi(x)$ if $\phi(x) \le a$, and $\phi_a(x) = a$ if $\phi(x) > a$. Then we can apply the above result to ϕ_a, and it remains to utilize the corresponding limit process (tending a to infinity). This establishes the inequality (3) for any $\phi \in \Phi$. Finally, in view of (3) and the independence of Z_1 and Z_2, the following relations are satisfied:

$$E(\phi(Z_1)) = \int_F \phi(x)\mu_1(dx) \le$$

$$\int_F (\int_F \phi(x+y)\mu_1(dx))\mu_2(dy) = E(\phi(Z_1 + Z_2)),$$

which yields the required inequality (2). Thus, the proof of Theorem 4 is completed.

Note that various statistical applications of the Anderson inequality are presented in many works and monographs (see, for instance, [96]).

EXERCISES

1. Let $X \in L_2(\Omega, \mathbf{R})$. Check that

$$E((X - E(X))^2) = min_{r \in \mathbf{R}} E((X - r)^2).$$

2. Let $X \in L_1(\Omega, \mathbf{R})$ and let G_X denote the distribution function of X. Verify that

$$E(|X - m|) = min_{r \in \mathbf{R}} E(|X - r|)$$

where $m \in \mathbf{R}$ is the median of G_X, i.e. m satisfies the inequalities

$$G_X(m-) \le 1/2 \le G_X(m).$$

3. Give a detailed proof of properties I - III of the Pettis integral.

4. Give a more precise version of Theorem 1 in the case when F is a Frechet space and C is a compact subset of F. Namely, demonstrate that, for any point $y \in F$, the following two assertions are equivalent:

a) $y \in cl(conv(C))$;
b) there exists a Radon probability measure μ on C such that

$$y = \int_C x\mu(dx).$$

In connection with the result of Exercise 4, let us mention that further interesting and important statements of this type can be found, e.g., in [31], [56] and [163].

5. Let us put:
$\Omega = \mathbf{N}$;
$S =$ the class of all subsets of Ω;
$P =$ the discrete probability measure on S defined by the formula

$$P(\{2n\}) = P(\{2n+1\}) = p_n \qquad (n \in \mathbf{N})$$

where $(p_n)_{n\in\mathbf{N}}$ is a countable family of strictly positive real numbers, such that

$$\sum_{n\in\mathbf{N}} p_n = 1/2.$$

Further, define a random element

$$X : \Omega \to c_0$$

by the formula

$$X(n) = (x_k^{(n)})_{k\in\mathbf{N}} \qquad (n \in \Omega = \mathbf{N}),$$

where, for all $j \in \mathbf{N}$, we put:

$$x_k^{(2j)} = 1/p_j \quad iff \quad k = 2j,$$

$$x_k^{(2j)} = 0 \quad iff \quad k \neq 2j,$$

$$x_k^{(2j+1)} = -(1/p_j) \quad iff \quad k = 2j+1,$$

$$x_k^{(2j+1)} = 0 \quad iff \quad k \neq 2j+1.$$

Check that:
a) X is not integrable in the Bochner sense;
b) X is integrable in the Pettis sense and

$$\int_\Omega X(\omega)P(d\omega) = 0;$$

c) X is not integrable in the Pettis sense on the set

$$\Omega' = \{0, 2, 4, ..., 2n, ...\} \subset \Omega.$$

6. Prove, for the Bochner integral, an analogue of the Lebesgue classical theorem on majorized convergence. Namely, let $(X_n)_{n \in \mathbf{N}}$ be a sequence of random elements in a separable Banach space $(F, ||\cdot||)$, such that

$$E(sup_{n \in \mathbf{N}} ||X_n||) < +\infty$$

and, for some random element X in this space,

$$P(\{\omega \ : \ lim_{n \to \infty} ||X_n(\omega) - X(\omega)|| = 0\}) = 1.$$

Then the equality

$$lim_{n \to \infty} E(X_n) = E(X)$$

holds true.

7. Let X be a Gaussian random element in a separable Banach space $(F, ||\cdot||)$. Show that there exists a real number $\varepsilon > 0$ for which

$$E(exp(\varepsilon ||X||^2)) < +\infty.$$

This result is stronger than the assertion of Theorem 2.

8. Give a detailed proof of properties 1 - 7 of conditional means.

9. Check that it is impossible to define conditional means via Pettis integral.

10. Show, under the assumptions of Theorem 4, that if a loss functional $\phi \in \Phi$ satisfies the condition

$$\phi(x) > 0 \qquad (x \in F, \ x \neq 0),$$

then $E(X|Y)$ is a unique optimal estimator of X (by watching Y) with respect to ϕ.

11. Extend Theorem 4 to the case when two jointly Gaussian elements X and Y are given, respectively, in some separable Banach spaces F_1 and F_2 (in general, $F_1 \neq F_2$).

12. Let F be a Banach space and let F^* be its conjugate space equipped with the topology $\sigma(F^*, F)$. Suppose that a linear form

$$u \ : \ F^* \to \mathbf{R}$$

is given. Prove that these three relations are equivalent:

1) u is continuous;

2) there exists a vector $x \in F$ such that

$$(\forall h \in F^*)(u(h) = h(x));$$

3) the restriction of u to the closed unit ball

$$B^* = \{h \in F^* \ : \ ||h|| \leq 1\}$$

is continuous (here B^* is equipped with the topology induced by $\sigma(F^*, F)$).

The main difficulty arises in connection with the implication 3) $\Rightarrow$ 2). Try to establish this implication in the following manner.

For any set $X \subset F$, denote

$$pol(X) = \{h \in F^* \ : \ (\forall x \in X)(|h(x)| \leq 1)\}.$$

The set $pol(X)$ is usually called the polar of X. Assuming that relation 3) holds and applying the compactness of all homothetic images of B^*, construct recursively a sequence $(X_n)_{n \in \mathbf{N}}$ of finite subsets of F, satisfying the conditions:

a) for each natural number n, the inclusion

$$pol(X_0 \cup X_1 \cup ... \cup X_n) \cap (n+1)B^* \subset u^{-1}(]-1, 1[)$$

is valid;

b) for each natural number $n \geq 1$ and for any element $x \in X_n$, we have $||x|| \leq 1/n$.

Further, let $(x_n)_{n \in \mathbf{N}}$ be an injective enumeration of all elements of the set $\cup\{X_n \ : \ n \in \mathbf{N}\}$. Check that

$$lim_{n \to \infty} x_n = 0.$$

Check also that if $h \in pol(\{x_n \ : \ n \in \mathbf{N}\})$, then

$$|u(h)| \leq 1.$$

Now, introduce a linear mapping

$$\Phi \ : \ F^* \to c_0$$

by putting

$$\Phi(h) = (h(x_0), h(x_1), ..., h(x_n), ...) \qquad (h \in F^*).$$

The range of Φ is a vector subspace of c_0 on which a linear functional Ψ can be defined by the formula

$$\Psi(\Phi(h)) = u(h) \qquad (h \in F^*).$$

Verify that the definition of Ψ is correct and that Ψ is continuous on its domain. According to the Hahn-Banach theorem, there exists an extension of Ψ to a continuous linear functional on c_0. We denote this extension by the same symbol Ψ. As shown in Section 5, the extended functional Ψ must be of the form

$$\Psi((t_n)_{n\in\mathbf{N}}) = \sum_{n\in\mathbf{N}} a_n t_n \qquad ((t_n)_{n\in\mathbf{N}} \in c_0),$$

where $(a_n)_{n\in\mathbf{N}}$ is some fixed element of the space l_1. Verify that

$$(\forall h \in F^*)(u(h) = h(x))$$

where $x \in F$ is defined by

$$x = \sum_{n\in\mathbf{N}} a_n x_n.$$

Finally, suppose that the original Banach space F is separable (hence B^* is compact and metrizable). In this situation, demonstrate that 3) is equivalent to the following relation:

3') for any sequence $(h_n)_{n\in\mathbf{N}} \subset F^*$ weakly converging to zero, we have

$$lim_{n\to\infty} u(h_n) = 0.$$

It was already mentioned in this section that the above result is due to Banach. More general versions of this result (for some classes of locally convex topological vector spaces) can be found in [29] and [179].

Appendix 1: Some properties of convex curves

Here we discuss some interesting properties of convex curves lying in the Euclidean plane $\mathbf{R}^2$. Namely, in our further considerations, we shall meet several purely geometric characteristics of such curves, which are important in various applications. We wish to warn the reader that the facts considered below are very elementary and their role is only to illustrate some typical arguments utilized in geometric probability and integral geometry. Extensive information from these domains of mathematics can be found, e.g., in the well-known monographs by Hadwiger and Santaló (see [85] and [178]).

Let us recall that a curve in $\mathbf{R}^2$ is defined, as a rule, by the formula

$$F(x, y) = 0$$

where F is some function of two variables, satisfying certain analytic assumptions which are implied by the geometric features of a given curve. Quite often we deal not only with a single curve, but also with a family of curves, defined by

$$F(x, y, \theta) = 0$$

where F is a function of three variables and θ is a parameter ranging over a fixed set $\Theta \subset \mathbf{R}$. In this case, we have a parametrized family of curves in $\mathbf{R}^2$. Frequently, the set Θ coincides with some subinterval of $\mathbf{R}$.

If F satisfies suitable analytic conditions, then, according to the classical theorem of differential geometry, there exists an envelope of this family of curves, and its equation can be obtained from the system

$$F(x, y, \theta) = 0, \quad F'_\theta(x, y, \theta) = 0,$$

by excluding a parameter θ. On the other hand, sometimes it is possible to express both variables x and y in terms of θ:

$$x = x(\theta), \quad y = y(\theta) \qquad (\theta \in \Theta),$$

which yields a parametric representation of the envelope.

Let us consider a very particular case when all curves from our family are straight lines in the plane. By using the so-called normal form of the equation of a straight line, we may represent each line from the family as

$$x cos(\phi) + y sin(\phi) - p(\phi) = 0$$

where a parameter $\phi \in [0, 2\pi[$ plays the role of θ, and

$$p \; : \; [0, 2\pi[\rightarrow [0, +\infty[$$

is a certain function (actually, for any ϕ, the value $p(\phi)$ coincides with the distance between the origin and our straight line).

Assuming the differentiability of p with respect to ϕ and taking the corresponding derivative, we easily obtain

$$-x sin(\phi) + y cos(\phi) - p'(\phi) = 0.$$

The two formulae above immediately yield one of the parametric representations of the envelope:

$$x(\phi) = p(\phi) cos(\phi) - p'(\phi) sin(\phi),$$

$$y(\phi) = p(\phi) sin(\phi) + p'(\phi) cos(\phi).$$

Moreover, for each $\phi \in [0, 2\pi[$, the straight line corresponding to the equation

$$x cos(\phi) + y sin(\phi) - p(\phi) = 0$$

turns out to be tangential to the envelope at its point $(x(\phi), y(\phi))$.

Furthermore, assuming the existence of a second derivative of our function p, we get

$$x'(\phi) = -(p(\phi) + p''(\phi)) sin(\phi),$$

$$y'(\phi) = (p(\phi) + p''(\phi)) cos(\phi).$$

Denoting by $s = s(\phi)$ the length of an arc of the envelope, we come to the formula

$$(s'(\phi))^2 = (p(\phi) + p''(\phi))^2.$$

Now, remembering the classical definitions of a curvature and of a curvature radius, we see that

$$k(\phi) = \frac{d\phi}{ds} = \frac{1}{p(\phi) + p''(\phi)},$$

$$r(\phi) = \frac{1}{|k(\phi)|} = |p(\phi) + p''(\phi)|,$$

where $k(\phi)$ (respectively, $r(\phi)$) denotes the curvature (respectively, the curvature radius) of our envelope at its point $(x(\phi), y(\phi))$.

Let us consider more thoroughly the question concerning the correct choice of a sign of $k(\phi)$ in the above-mentioned formula. If the functions

$$\phi \rightarrow x(\phi), \qquad \phi \rightarrow y(\phi) \qquad (\phi \in [0, 2\pi[)$$

are assumed to be twice differentiable, then, in virtue of the well-known theorem from classical differential geometry, we have

$$k(\phi) = \frac{x'(\phi)y''(\phi) - x''(\phi)y'(\phi)}{((x'(\phi))^2 + (y'(\phi))^2)^{3/2}}.$$

It can directly be seen from this equality that the sign of $k(\phi)$ coincides with the sign of

$$x'(\phi)y''(\phi) - x''(\phi)y'(\phi),$$

i.e. with the sign of $p(\phi)+p''(\phi)$. Further, as known from differential geometry, a given curve lying in the plane is convex if and only if its curvature is positive. We thus conclude that the convexity of our envelope is equivalent to the relation

$$p(\phi) + p''(\phi) \geq 0 \qquad (\phi \in [0, 2\pi[).$$

Let us recall that, in our preceding considerations, we supposed that the function p satisfies some smoothness conditions. But, actually, these conditions turn out to be superfluous because each convex curve in $\mathbf{R}^2$ admits rather good approximation by smooth strongly convex curves (cf. Exercise 10 from Section 2). In this connection, let us also recall that any convex curve in $\mathbf{R}^2$ is twice differentiable at almost all of its points. This fact is an immediate consequence of the classical Lebesgue theorem stating the differentiability almost everywhere of any monotone function acting from $\mathbf{R}$ into itself (see, e.g., [154] or [177]).

The first nontrivial result which can be deduced from the simple considerations presented above is the theorem on the existence of four top-points of an oval lying in $\mathbf{R}^2$. This old result is due to Muckhopadhyaya. It can be formulated as follows:

Theorem 1. *For any oval in the plane $\mathbf{R}^2$ (i.e. for any closed convex curve in $\mathbf{R}^2$ whose coordinate functions possess continuous second derivatives), there exist at least four top-points (i.e. there exist at least four points of the curve, such that the curvature k has a local extremum at each of them).*

Proof. Since $k = 1/r$, where r denotes the curvature radius of our oval, it suffices to demonstrate an analogous assertion for r. We may suppose without loss of generality that r is not a constant function (otherwise, there is nothing to prove). Further, since r is continuous, it is evident that at least two top-points must exist: at one of them r has a global maximum, and at the second one r has a global minimum. On the other hand, between any two local maxima (minima) there always lies at least one minimum (maximum). So, the total number of top-points is either equal to 2, or is greater than or equal to 4. Let us establish that the first case is impossible. Indeed, taking into account the formulae above, we may write

$$\int_0^{2\pi} r(\phi)sin(\phi)d\phi = -\int_0^{2\pi} x'(\phi)d\phi = -x(2\pi) + x(0) = 0,$$

because our curve is closed. Moreover, it will be convenient to assume in the sequel that r is a periodic function (with period 2π) defined on the whole real line $\mathbf{R}$.

Suppose now that our oval has only two top-points corresponding to the global maximum and global minimum. We may put, without loss of generality, that:

$(x(-\pi/2), y(-\pi/2))$ = the point of a global minimum of r;

$(x(\pi/2), y(\pi/2))$ = the point of a global maximum of r.

Then the function r turns out to be strictly increasing on the segment $[-\pi/2, \pi/2]$ and strictly decreasing on the segment $[\pi/2, 3\pi/2]$. This yields, in particular, that

$$r(\phi) > r(-\phi) \qquad (\phi \in [0, \pi]).$$

At the same time, we have

$$0 = \int_0^{2\pi} r(\phi)sin(\phi)d\phi = \int_{-\pi}^{\pi} r(\phi)sin(\phi)d\phi =$$

$$\int_{-\pi}^{0} r(\phi)sin(\phi)d\phi + \int_0^{\pi} r(\phi)sin(\phi)d\phi = \int_0^{\pi} (r(\phi) - r(-\phi))sin(\phi)d\phi > 0,$$

which is impossible. The contradiction obtained shows us that there are at least four top-points of the given oval, and the theorem has thus been proved.

Some generalizations and extensions of this beautiful theorem can be found in [16] and [18].

Let us mention one useful fact implied by the formula

$$s'(\phi) = p(\phi) + p''(\phi) \qquad (\phi \in [0, 2\pi[)$$

for a closed convex curve lying in the plane $\mathbf{R}^2$. Integrating both sides of this formula and taking into account the periodicity of p (since our curve is closed), we get

$$L = \int_0^{2\pi} s'(\phi)d\phi = \int_0^{2\pi} p(\phi)d\phi + p'(\phi)|_0^{2\pi} = \int_0^{2\pi} p(\phi)d\phi,$$

where L denotes the entire length of the curve.

Now, notice that if the origin of $\mathbf{R}^2$ lies in the interior of the convex hull of our curve, then the function p coincides with the Minkowski support function h for this convex hull (see Section 1). We thus conclude that the length L can be expressed in the integral form by means of the support function:

$$L = \int_0^{2\pi} h(\phi)d\phi.$$

The previous argument was based on the assumption of the differentiability of h. But, applying the approximation theorem, we may assert that the latter formula holds true for any closed convex curve on the plane.

In a similar way, we can express the area of the figure determined by our curve (in terms of h and its derivatives). Namely, utilizing the classical formula from analysis, we easily obtain

$$S = (1/2)\int_0^{2\pi} (x(\phi)y'(\phi) - x'(\phi)y(\phi))d\phi =$$

$$(1/2)\int_0^{2\pi} h(\phi)(h(\phi) + h''(\phi))d\phi$$

where S denotes the area of the figure. The integration by parts reduces the last integral to the following one:

$$S = (1/2)\int_0^{2\pi} (h^2(\phi) - (h'(\phi))^2)d\phi,$$

i.e. the area S is expressed in terms of h and its first derivative.

Let us demonstrate one simple application of the formulae presented above. We recall that a closed convex curve in $\mathbf{R}^2$ is a curve of constant width $d > 0$ if the orthogonal projection of the curve on any straight line (lying in $\mathbf{R}^2$) is a segment whose length equals d. Similarly, a two-dimensional compact convex subset K of $\mathbf{R}^2$ has constant width d if the orthogonal projection of K on any straight line in $\mathbf{R}^2$ is a segment with length d. Evidently, K is of constant width d if and only if its boundary is of the same constant width. Also, it can easily be observed that K has constant width d if and only if

$$h(\phi) + h(\phi + \pi) = d$$

for all $\phi \in [0, 2\pi[$ (where, as usual, h denotes the Minkowski support function for K, canonically extended on the whole real line $\mathbf{R}$). Integrating this equality along the segment $[0, 2\pi]$ and taking into account the periodicity of h, we get

$$2L = 2\pi d, \quad L = \pi d,$$

where L is the length of the boundary of K. Thus, we conclude that all closed convex curves of constant width d have constant length equal to πd. For areas, an analogous assertion is not true (the reader can check this fact by constructing simple examples). In connection with curves of constant width, let us point out the following nontrivial statement: for any set $X \subset \mathbf{R}^2$ with diameter d, there exists a compact convex set $K \subset \mathbf{R}^2$ of constant width d, containing X (obviously, K has the same diameter d). In other words, each subset of $\mathbf{R}^2$ having diameter d can be isometrically embedded into some subset of $\mathbf{R}^2$ having constant width d. Notice that this interesting result remains true for sets of diameter d lying in the n-dimensional Euclidean space $\mathbf{R}^n$ (see, e.g., [21]).

Let us return to the family of all straight lines in the plane and to its various subfamilies. As mentioned earlier, any such a line can be given by the equation in the normal form:

$$x cos(\phi) + y sin(\phi) - p = 0,$$

where $\phi \in [0, 2\pi[$ and $0 \leq p < +\infty$. Consider the product set

$$[0, 2\pi[\times [0, +\infty[.$$

This set is equipped with the two-dimensional Lebesgue measure induced by the standard Lebesgue measure on $\mathbf{R}^2$. Further, the canonical correspondence between the above product set and the family of all straight lines in $\mathbf{R}^2$ transfers the Lebesgue measure to this family (let us remark that the correspondence just mentioned is not bijective but is "almost" bijective). We shall denote the obtained measure by the symbol μ. Thus, μ is a certain σ-finite measure defined on the σ-algebra of all Borel subsets of this family (we suppose that the family is equipped with its natural topology). Sometimes, it is more convenient to deal with the completion of μ; for the sake of simplicity, we shall denote the completed measure by the same symbol μ. Note that the group of all motions (i.e. isometric transformations) of $\mathbf{R}^2$ may also be regarded as a group of transformations of the family of all straight lines lying in $\mathbf{R}^2$. Hence, in virtue of the invariance of the standard two-dimensional Lebesgue measure under the group of all motions of the plane, we can assert that μ is an invariant measure with respect to the same group. This observation is important, of course, and implies numerous consequences. However, we do not need the invariance of μ in our further considerations (briefly speaking, this property of μ does not play an essential role for us).

Let K be a compact convex set in $\mathbf{R}^2$ such that $0 \in int(K)$. It is not hard to see that the set

$$Z = Z(K) = \{l \ : \ l \cap K \neq \emptyset\}$$

of all those straight lines l in $\mathbf{R}^2$ which intersect K is μ-measurable, and our purpose is to find its μ-measure. In order to do this, let us first remark that the line in $\mathbf{R}^2$ satisfying the equation

$$x cos(\phi) + y sin(\phi) - p = 0$$

has nonempty intersection with K if and only if $p \leq h(\phi)$ where h denotes the Minkowski support function for K. Applying this evident fact, we infer

$$\mu(Z) = \int\int_{Z^*} dp d\phi = \int_0^{2\pi} h(\phi) d\phi = L$$

where Z^* is the image (in $\mathbf{R}^2$) of the set Z, with respect to the canonical correspondence, and $L = L(bd(K))$ is the length of the boundary of K.

This simple but important formula was discovered by Crofton. It immediately implies the following result of geometric-probabilistic character.

Theorem 2. *Let K_1 and K_2 be any two compact convex subsets of the plane, such that $K_1 \subset K_2$. Then the probability of the event that a random straight line in the plane intersects K_1, under the condition that this line has nonempty intersection with K_2, is equal to L_1/L_2 where L_i $(i = 1, 2)$ denotes the length of the boundary of K_i.*

We thus see that Theorem 2 is a generalization of the well-known statement from elementary geometry, according to which L_1 is always less than or equal to L_2.

Let now K be an arbitrary Borel subset of $\mathbf{R}^2$ and let l be a straight line in $\mathbf{R}^2$. Suppose that l is defined by the equation

$$x cos(\phi) + y sin(\phi) - p = 0.$$

In this case, we shall write $l = l(\phi, p)$. Also, denote $\lambda_K(\phi, p) = \lambda(K \cap l)$ where λ is the standard one-dimensional Lebesgue measure on l. Further, consider the integral

$$\int\int_{Z^*} \lambda_K(\phi, p) d\phi dp$$

where, as above, the set $Z = Z(K)$ is given by

$$Z = \{l \ : \ l \cap K \neq \emptyset\},$$

and Z^* is the image of Z (with respect to the canonical correspondence). Applying the classical Fubini theorem, we easily infer that, for each ϕ from the interval $[0, \pi[$, the sum

$$\int \lambda_K(\phi, p) dp + \int \lambda_K(\phi + \pi, p) dp$$

coincides with $\lambda_2(K)$ where λ_2 denotes the standard Lebesgue measure on $\mathbf{R}^2$. So, we may write

$$\int\int_{Z^*} \lambda_K(\phi, p) d\phi dp = \pi \lambda_2(K).$$

Sometimes, it is more convenient to deal with oriented straight lines of the plane $\mathbf{R}^2$. In this case $\mu(Z) = 2\pi\lambda_2(K)$, which is evident from the geometrical point of view.

Let us return to convex curves lying in $\mathbf{R}^2$. Let P be a closed convex piecewise differentiable curve in the plane, having length $L = L(P)$. We consider the equation of this curve in the so-called natural parametric form:

$$x = x(s), \quad y = y(s) \qquad (0 \leq s \leq L),$$

where s is the natural parameter of P, i.e. the length of an arc of P. Suppose now that some oriented straight line l lying in $\mathbf{R}^2$ intersects our curve in a point $(x(s), y(s))$ (and, in addition, $(x(s), y(s))$ is the first point of l at which l meets P). We denote by θ the angle between l and the tangential vector $(x'(s), y'(s))$. Notice that θ is determined uniquely (since l is oriented), and we may assume that $0 \leq \theta \leq \pi$. Thus, l can be treated as a certain function of (s, θ), i.e. we can put $l = l(s, \theta)$. In other words, s and θ will be considered as coordinates of l. But, as we know, the same (but non-oriented) line l is defined by its normal equation

$$x cos(\phi) + y sin(\phi) - p = 0$$

where ϕ and p are also regarded as coordinates of l. Consequently, on the set $Z = \{l \ : \ l \cap P \neq \emptyset\}$, we have the following mapping:

$$p = p(s, \theta), \quad \phi = \phi(s, \theta).$$

Our goal is to calculate the Jacobian of this mapping. It is not hard to see that

$$\theta = \phi - \tau \pm \pi/2$$

where τ denotes the angle between the unit vector $(x'(s), y'(s))$ and the axis of abscissae. Equivalently, the angle $\tau = \tau(s)$ is determined by the formulae:

$$x'(s) = cos(\tau), \quad y'(s) = sin(\tau).$$

Further, we have the relation

$$p(s, \theta) = x(s)cos(\phi(s, \theta)) + y(s)sin(\phi(s, \theta))$$

for all s (except finitely many of them) and for all θ satisfying the inequalities

$$0 \le s \le L, \quad 0 \le \theta \le \pi.$$

Taking the full differential, we get

$$dp = cos(\phi)dx + sin(\phi)dy + (ycos(\phi) - xsin(\phi))d\phi.$$

Since $x'(s) = cos(\tau)$ and $y'(s) = sin(\tau)$, we may write

$$dx = cos(\tau)ds, \quad dy = sin(\tau)ds$$

whence it follows that

$$dp = cos(\phi - \tau)ds + (ycos(\phi) - xsin(\phi))d\phi.$$

Here we have the equality of two differential forms. Let us take the outer product of dp and $d\phi$. Then, in virtue of the relation

$$d\phi \wedge d\phi = 0,$$

we come to the equality

$$dpd\phi = cos(\phi - \tau)dsd\phi.$$

In addition to this, the formula $\theta = \phi - \tau \pm \pi/2$ implies

$$d\theta = d\phi - \tau'(s)ds.$$

Taking the outer product of ds and $d\theta$, we get

$$ds d\theta = ds d\phi$$

because $ds \wedge ds = 0$. Finally, we obtain

$$dp d\phi = cos(\phi - \tau) ds d\theta,$$

i.e. the Jacobian of our mapping is identical with $cos(\phi - \tau)$. But, utilizing the formula

$$|cos(\phi - \tau)| = sin(\theta),$$

we are able to express the modulus of the Jacobian in a simplified form. Actually, we may write

$$d\phi dp = dp d\phi = sin(\theta) ds d\theta.$$

This formula enables us to calculate $\mu(Z)$. Namely, we get

$$\mu(Z) = 2 \int\int_{Z^*} dp d\phi = \int_0^L ds \int_0^\pi sin(\theta) d\theta = 2L.$$

Let us stress that the argument above is applicable only to closed convex piecewise differentiable curves in $\mathbf{R}^2$. However, it can be demonstrated that the result remains true for any closed convex curve P lying in the plane (and for much more general curves). We suggest the reader to prove this generalized statement by utilizing the approximation theorem. Also, we wish to note that if P is a closed convex curve in $\mathbf{R}^2$, then the result just mentioned can be reduced to the formula $\mu(Z) = L$ which was established earlier with the aid of the Minkowski support function. Indeed, the coefficient 2 at L appears here because the oriented straight lines of the plane are under consideration. In this connection, it is reasonable to recall that the orientation of a straight line in $\mathbf{R}^2$ is necessary for introducing the coordinates s and θ.

We want to conclude this Appendix with one elegant proof of the isoperimetric property of a circle. This proof is due to Blaschke and provides a good illustration of the usefulness of the approach described above. We return to the formula $\mu(Z) = 2\pi\lambda_2(K)$ where K is a two-dimensional compact subset of $\mathbf{R}^2$ and Z is the family of all those oriented lines in the plane which intersect K. Let us denote by P the boundary of K and suppose that P is homeomorphic to a circumference and has finite length $L = L(P)$.

Theorem 3. *Assuming that L is fixed, the value $\lambda_2(K)$ is maximal if and only if the curve P is a circumference.*

Proof. First of all, we may consider (without loss of generality) only the case when our compact set K is convex. Indeed, suppose for a moment that K is not convex. Then

$$conv(K) \neq K, \quad K \subset conv(K),$$

and it is easy to see that

$$\lambda_2(conv(K)) > \lambda_2(K), \quad L(bd(conv(K))) \leq L(P),$$

so the value $\lambda_2(K)$ cannot be maximal. Thus, in our further considerations, we may assume that K is convex. Let

$$x = x(s), \quad y = y(s) \qquad (0 \leq s \leq L)$$

be the natural equation of the curve P. As earlier, for each $s \in [0, L]$ and for any oriented straight line l lying on the plane and intersecting P in the point $(x(s), y(s))$, we denote by θ the angle between l and the unit vector $(x'(s), y'(s))$. Note that, since P is convex, $x'(s)$ and $y'(s)$ exist for all values $s \in [0, L]$ except countably many of them. Also, it is clear that if $x'(s)$ and $y'(s)$ exist, then the straight line passing through the points $(x(s), y(s))$ and $(x(s) + x'(s), y(s) + y'(s))$ is support for K. Let us denote by $\lambda(s, \theta)$ the length of the chord $K \cap l$ and consider the integral

$$I = \int\int\int\int (\lambda(s,\theta)sin(\theta') - \lambda(s',\theta')sin(\theta))^2 ds ds' d\theta d\theta'$$

along the four-dimensional parallelepiped

$$\{(s, s', \theta, \theta') \ : \ 0 \leq s \leq L, \ 0 \leq s' \leq L, \ 0 \leq \theta \leq \pi, \ 0 \leq \theta' \leq \pi\}.$$

Obviously, this integral is greater than or equal to 0. Our goal is to calculate it. For this purpose, note first that

$$\int_0^\pi sin^2(\theta)d\theta = \int_0^\pi sin^2(\theta')d\theta' = \pi/2$$

and, for almost all s and s' from $[0, L]$, we have

$$\int_0^\pi (\lambda(s,\theta))^2 d\theta = \int_0^\pi (\lambda(s',\theta'))^2 d\theta' = 2\lambda_2(K).$$

The last formula can easily be obtained by using the polar system of coordinates, whose origin coincides with $(x(s), y(s))$ (respectively, with $(x(s'), y(s')))$ and whose axis is tangential to P at $(x(s), y(s))$ (respectively, at $(x(s'), y(s')))$.

Taking into account these formulae, we come to the equality

$$I = 2\pi L^2(P)\lambda_2(K) - 2\int\int \lambda(s,\theta)sin(\theta)dsd\theta \int\int \lambda(s',\theta')sin(\theta')ds'd\theta'.$$

On the other hand, we have

$$\int\int \lambda(s,\theta)sin(\theta)dsd\theta = \int\int \lambda(s',\theta')sin(\theta')ds'd\theta'$$

$$= 2\int\int \lambda_K(\phi,p)d\phi dp = 2\pi\lambda_2(K).$$

This implies

$$I = 2\pi\lambda_2(K)(L^2(P) - 4\pi\lambda_2(K)) \geq 0,$$

which trivially yields the desired isoperimetric inequality. It remains to specify the case when the isoperimetric inequality is reduced to the equality. This can be done without any difficulty. Indeed, the relation $I = 0$ is equivalent to

$$\lambda(s,\theta)sin(\theta') - \lambda(s',\theta')sin(\theta) = 0$$

for all s and s' from $[0, L]$ (except countably many of them) and for all θ and θ' from $[0, \pi]$. Geometrically, the latter relation means that each chord of K forms equal angles with the tangential straight lines to P passing through the end-points of the chord. Now, a simple geometric argument (left to the reader) shows that, in such a case, P must be a circumference, and the proof of the theorem is finished.

As mentioned earlier, many other proofs of the isoperimetric inequality (with its generalizations to the case of a high-dimensional Euclidean space) may be found in [16], [18], [47], [72], [85] and [178].

Appendix 2: Convex sets and number theory

We have already mentioned in the Preface that Minkowski (one of the founders of the theory of convex sets) successfully applied his theory to various problems concerning deep properties of natural numbers. In this connection, it must also be noted that the methods created by Minkowski became basic for further development of the so-called geometric number theory in which purely arithmetic (or algebraic) properties of natural numbers are studied from the geometric viewpoint.

Here we present direct applications of his methods to some classical questions from number theory. For more serious information about this topic, we refer the reader to [26], [49], [50], [88] and [151].

First, let us introduce several auxiliary notions which are useful not only in number theory, but also in other domains of mathematics.

Let E be a set and let G be some group of transformations of E. The pair (E, G) is usually called a space with a transformation group. Such spaces play the fundamental role in the contemporary geometry and topology. Moreover, from the modern point of view, any geometry over E can be described in terms of invariants of an appropriate transformation group G acting in E.

We need the following two definitions.

Definition 1. A subset X of E is G-thick in E if

$$\cup\{g(X) \ : \ g \in G\} = E.$$

Definition 2. A subset Y of E is G-thin in E if

$$(\forall g \in G)(\forall h \in G)(g \neq h \Rightarrow g(Y) \cap h(Y) = \emptyset).$$

Let us remark that a G-thick subset of E can be G-thin as well. For example, if $E = G = \mathbf{R}$, then all one-element subsets of E are G-thick and G-thin simultaneously.

Let us present two simple propositions concerning G-thick and G-thin sets in E.

Lemma 1. *Suppose that (E, G) is a space with a transformation group and μ is a σ-finite G-invariant measure on E. Let X be some μ-measurable G-thick subset of E and let Y be some μ-measurable G-thin subset of E. Then the inequality*

$$\mu(Y) \leq \mu(X)$$

holds true.

Proof. If our group G is uncountable, then the disjoint family of sets $\{g(Y) : g \in G\}$ is uncountable, too, and taking into account the G-invariance and the σ-finiteness of μ, we easily get $\mu(Y) = 0$ and hence $\mu(Y) \leq \mu(X)$. Thus, in the case of an uncountable G, the assertion is trivially valid.

Assume now that G is countable. Then we may write

$$\mu(Y) \leq \sum_{g \in G} \mu(Y \cap g(X)) = \sum_{g \in G} \mu(g^{-1}(Y) \cap X) \leq \mu(X),$$

because the family $(g^{-1}(Y) \cap X)_{g \in G}$ is countable and consists of pairwise disjoint μ-measurable subsets of X.

Note that, in the second case, we applied only the countable additivity and G-invariance of μ (the σ-finiteness of μ was not necessary).

We see that the required inequality holds true in the both cases, and Lemma 1 has thus been proved.

Lemma 2. *Let (E, G) be a space with a transformation group, let H be a subgroup of G and let $(Hg_i)_{i \in I}$ denote the partition of G into equivalence classes modulo H. Then the following assertions are satisfied:*

1) if a set $X \subset E$ is G-thick in E, then the set $\cup\{g_i(X) : i \in I\}$ is H-thick in E;

2) if a set $Y \subset E$ is G-thin in E, then the set $\cup\{g_i(Y) : i \in I\}$ is H-thin in E.

The proof is not hard and is left to the reader.

Now, we are able to demonstrate the classical Minkowski theorem (in a slightly more general formulation).

Theorem 1. *Let H be a discrete group of translations of the Euclidean space $\mathbf{R}^n$, where $n \geq 1$, and let T be a compact convex body in this space, symmetric with respect to the origin. Furthermore, let λ_n denote the standard n-dimensional Lebesgue measure on $\mathbf{R}^n$ and let X be an arbitrary λ_n-measurable H-thick subset of $\mathbf{R}^n$. If the inequality*

$$\lambda_n(T) \geq 2^n \lambda_n(X)$$

is true, then there exists a nonzero vector $h \in H \cap T$.

Proof. First, let us consider the case when

$$\lambda_n(T) > 2^n \lambda_n(X).$$

In this case, we may write

$$\lambda_n(T/2) > \lambda_n(X).$$

Applying Lemma 1 to the sets $T/2$ and X, we infer that $T/2$ cannot be H-thin in $\mathbf{R}^n$. Consequently, there exist two distinct vectors $h' \in H$ and $h'' \in H$ such that

$$(h' + T/2) \cap (h'' + T/2) \neq \emptyset.$$

This implies that

$$h' - h'' \in (T/2) + (-T/2) = T/2 + T/2 = T,$$

because T is convex and symmetric with respect to the origin. Thus, the nonzero vector $h' - h''$ lies in $H \cap T$ and is the required one.

Suppose now that

$$\lambda_n(T) = 2^n \lambda_n(X)$$

or, equivalently,

$$\lambda_n(T/2) = \lambda_n(X).$$

For a natural number $k > 0$, consider the closed $(1/k)$-neighbourhood of $T/2$ which is also convex and symmetric with respect to the origin. Denoting this neighbourhood by T_k and applying the result obtained above to T_k, we get a nonzero vector $h_k \in H \cap T_k$. Now, since our group H is discrete (and, consequently, closed in $\mathbf{R}^n$), it is easy to see that there exists a vector $h \in H$ such that $h = h_k$ for infinitely many numbers k. Obviously,

$$h \neq 0, \quad h \in H \cap T,$$

which completes the proof.

Let us turn our attention to some immediate consequences of Theorem 1 in number theory.

Let a, b, c be integers and let $c \neq 0$. As usual, we write

$$a \equiv b \ (mod(c))$$

if c is a divisor of the difference $a - b$. Recall that $\equiv$ yields an equivalence relation in the set $\mathbf{Z}$ of all integers.

Fix any two integers a_1 and a_2 and consider a linear form

$$u : \mathbf{R}^2 \to \mathbf{R}$$

on the Euclidean plane, defined by

$$x \to a_1x_1 + a_2x_2 \qquad (x = (x_1, x_2) \in \mathbf{R}^2).$$

The following result concerning this form is due to Thue.

Theorem 2. *For any natural number $p \geq 1$, there exist integers z_1 and z_2 such that:*

1) $|z_1| + |z_2| > 0$, i.e the vector $z = (z_1, z_2) \in \mathbf{R}^2$ is nonzero;

2) $|z_1| < p^{1/2}$ and $|z_2| < p^{1/2}$;

3) $u(z_1, z_2) = a_1z_1 + a_2z_2 \equiv 0$ $(mod(p))$.

Proof. Let us denote

$$T = \{(x_1, x_2) \in \mathbf{R}^2 \ : \ |x_1| \leq p^{1/2}, \ |x_2| \leq p^{1/2}\}.$$

Clearly, T is a square whose centre coincides with the origin of the plane. In particular, T is convex and symmetric with respect to the origin. Moreover, we have

$$\lambda_2(T) = 4p$$

where λ_2 is the standard two-dimensional Lebesgue measure on $\mathbf{R}^2$. Further, we put:

G = the group of all those translations of $\mathbf{R}^2$ whose both coordinates are integers;

H = the subgroup of G consisting of all those elements $(x_1, x_2) \in G$ for which

$$u(x_1, x_2) = a_1x_1 + a_2x_2 \equiv 0 \ (mod(p)).$$

It can directly be checked that

$$(G : H) \leq p$$

where $(G : H)$ denotes, as usual, the index of H in G, i.e. $(G : H)$ is the number of equivalence classes in G modulo H. Now, let $(Hg_i)_{1 \leq i \leq k}$ be the partition of G into equivalence classes modulo H. As mentioned above, we have the inequality $k \leq p$. According to Lemma 2, the set

$$X = \cup_{1 \leq i \leq k} \, g_i([0, 1]^2)$$

is H-thick in the plane $\mathbf{R}^2$. In addition to this, we may write

$$\lambda_2(T) = 4p \geq 4k \geq 2^2\lambda_2(X).$$

So, by applying the Minkowski theorem, there exists a nonzero vector

$$z = (z_1, z_2) \in H \cap T.$$

Obviously, the coordinates z_1 and z_2 of this vector are the required integers.
This completes the proof of Theorem 2.

We suggest the reader to formulate and prove an appropriate analogue of Theorem 2 for the n-dimensional Euclidean space $\mathbf{R}^n$.

For our further purposes, we need a purely technical lemma. Actually, it is due to Euler.

Lemma 3. *Let x_1, x_2, x_3, x_4, y_1, y_2, y_3, y_4 be some elements of a commutative ring. Then the following equality holds true:*

$$(x_1^2 + x_2^2 + x_3^2 + x_4^2)(y_1^2 + y_2^2 + y_3^2 + y_4^2) = z_1^2 + z_2^2 + z_3^2 + z_4^2$$

where

$$z_1 = x_1y_1 + x_2y_2 + x_3y_3 + x_4y_4,$$

$$z_2 = x_1y_2 - x_2y_1 + x_3y_4 - x_4y_3,$$

$$z_3 = x_1y_3 - x_3y_1 + x_4y_2 - x_2y_4,$$

$$z_4 = x_1y_4 - x_4y_1 + x_2y_3 - x_3y_2.$$

The proof of this equality is reduced to its straightforward checking. From the geometric point of view (more precisely, from the point of view of classical theory of quaternions), the equality simply indicates that the norm of the product of any two quaternions coincides with the product of their norms.

Lemma 4. *Let p be an arbitrary prime natural number. Then there exist two natural numbers a and b such that*

$$a^2 + b^2 + 1 \equiv 0 \ (mod(p)).$$

Proof. If $p = 2$, then the assertion is trivial. Suppose now that $p > 2$ and consider the following two sequences of integers:

$$0^2,\ 1^2,\ \dots,\ m^2,\ \dots,\ ((p-1)/2)^2,$$

$$-1 - 0^2,\ -1 - 1^2,\ \dots,\ -1 - m^2,\ \dots,\ -1 - ((p-1)/2)^2.$$

Taking any two distinct terms m^2 and l^2 from the first sequence, we see that their difference

$$m^2 - l^2 = (m + l)(m - l)$$

is not divisible by the prime p because

$$0 < m + l < p, \quad 0 < |m - l| < p.$$

Thus, if we divide by p all terms of the first sequence, we must obtain pairwise distinct remainders. An analogous circumstance may be observed for the second sequence. But, the total amount of integers in both sequences equals $p + 1$, i.e. is strictly greater than p. Hence there exist a member a^2 from the first sequence and a member $-1 - b^2$ from the second one, such that the difference

$$a^2 - (-1 - b^2) = a^2 + b^2 + 1$$

is divisible by p, i.e.

$$a^2 + b^2 + 1 \equiv 0 \ (mod(p)).$$

This completes the proof of the lemma.

Lemma 5. *The volume of the four-dimensional Euclidean unit ball is equal to* $\pi^2/2$.

Proof. Here the general formula can be applied:

$$\lambda_n(B_n) = \frac{\pi^{n/2}}{\Gamma((n/2) + 1)},$$

where λ_n denotes the usual Lebesgue measure on the n-dimensional Euclidean space and B_n is the unit ball in this space. The above-mentioned formula is well known from the standard course of mathematical analysis. But, for our purpose, we do not need the whole power of this formula. Indeed, remembering its elementary analogue for the three-dimensional ball $B_3(0, r)$ with centre 0 and radius r, namely,

$$\lambda_3(B_3(0, r)) = (4/3)\pi r^3,$$

and applying the Fubini theorem, we obtain

$$\lambda_4(B_4) = \int_{-1}^{1} (4/3)\pi(1 - x^2)^{3/2} dx.$$

An easy calculation of this integral leads to the required equality

$$\lambda_4(B_4) = \pi^2/2.$$

Now, we are able to demonstrate the following old result in number theory, due to Lagrange.

Theorem 3. *Every natural number can be represented as the sum of four squares of integers.*

Proof. According to the fundamental theorem of arithmetic, each natural number is a product of some primes. So, in virtue of Lemma 3, it suffices to show that any prime number can be represented as the sum of four squares of integers. In order to do this, fix a prime p and, by using Lemma 4, find two integers a and b for which

$$a^2 + b^2 + 1 \equiv 0 \ (mod(p)).$$

Let ε be some strictly positive real number, the precise value of which will not play an essential rôle in our further argument. In the space $\mathbf{R}^4$, consider the closed ball T with centre 0 and radius

$$r = (4 - \varepsilon)^{1/4} p^{1/2}.$$

Of course, we assume that $\varepsilon < 4$. Denote by G the group of all those translations of $\mathbf{R}^4$ whose coordinates are integers. Obviously, this group is discrete in $\mathbf{R}^4$. Define two linear forms u_1 and u_2 on $\mathbf{R}^4$ by putting

$$u_1(x) = u_1(x_1, x_2, x_3, x_4) = ax_1 + bx_2 - x_3,$$

$$u_2(x) = u_2(x_1, x_2, x_3, x_4) = bx_1 - ax_2 - x_4,$$

for all $x = (x_1, x_2, x_3, x_4)$ from $\mathbf{R}^4$. In addition to this, define a subgroup H of G consisting of all those elements $x \in G$ for which

$$u_1(x) \equiv 0 \ (mod(p)), \qquad u_2(x) \equiv 0 \ (mod(p)).$$

It is not hard to verify that

$$(G : H) \leq p^2.$$

Now, consider the partition $(Hg_i)_{1 \leq i \leq k}$ of G into equivalence classes modulo H. It can easily be seen that the inequality $k \leq p^2$ holds true. Putting

$$X = \cup_{1 \leq i \leq k} \ g_i([0,1]^4)$$

and applying Lemma 2, we infer that X is an H-thick subset of the space $\mathbf{R}^4$. Also, according to Lemma 5, we may write

$$\lambda_4(T) = (\pi^2/2)(4-\varepsilon)p^2.$$

Hence, taking ε sufficiently small, we obtain

$$\lambda_4(T) \geq (\pi^2/2)(4-\varepsilon)k > 2^4\lambda_4(X).$$

Consequently, in virtue of the Minkowski theorem, there exists a nonzero vector

$$z = (z_1, z_2, z_3, z_4) \in H \cap T.$$

For such z, we obviously have

$$0 < z_1^2 + z_2^2 + z_3^2 + z_4^2 \leq (4-\varepsilon)^{1/2}p < 2p.$$

Besides, the relations

$$(az_1 + bz_2)^2 \equiv z_3^2 \ (mod(p)),$$

$$(bz_1 - az_2)^2 \equiv z_4^2 \ (mod(p))$$

are valid in view of the definition of H. From these two relations, remembering that

$$a^2 + b^2 \equiv -1 \ (mod(p)),$$

we immediately come to the relation

$$z_1^2 + z_2^2 + z_3^2 + z_4^2 \equiv 0 \ (mod(p))$$

which means that our prime p divides the sum $z_1^2 + z_2^2 + z_3^2 + z_4^2$. But, as mentioned above,

$$z_1^2 + z_2^2 + z_3^2 + z_4^2 < 2p,$$

so we can conclude that

$$z_1^2 + z_2^2 + z_3^2 + z_4^2 = p.$$

The Lagrange theorem has thus been proved.

Appendix 3: Measurability of cardinals

In Section 8, the theorem was proved concerning the existence of separable supports of Borel probability measures given on metrizable spaces whose weights are not real-valued measurable. Here we wish to discuss (briefly, of course) some problems closely connected with the existence of measurable cardinals. Also, we demonstrate the usefulness of these questions in related domains of mathematics. For more detailed information about this interesting topic, we refer the reader to the works [11], [78], [89], [99], [112], [130], [132], [181], [185] and [194].

First of all, we want to recall the corresponding definitions.

Let E be a set and let μ be a measure on some σ-algebra of subsets of E. We shall say that μ is diffused (or continuous) if, for each $e \in E$, the one-element set $\{e\}$ belongs to $dom(\mu)$ and $\mu(\{e\}) = 0$.

Definition 1. A cardinal number a is real-valued measurable if there exists a diffused probability measure whose domain coincides with the σ-algebra of all subsets of a.

Definition 2. A cardinal number a is two-valued measurable if there exists a two-valued diffused probability measure whose domain coincides with the σ-algebra of all subsets of a.

In virtue of these definitions, we immediately get:

(1) if a cardinal a is two-valued measurable, then it is real-valued measurable as well;

(2) if a cardinal a is two-valued (real-valued) measurable and a cardinal b is such that $a \leq b$, then b is two-valued (real-valued) measurable.

We thus see that the class of all those cardinals which are not two-valued (real-valued) measurable forms an initial interval in the class of all cardinals.

For our further purposes, we need also the notions of a regular cardinal number and of a singular cardinal number.

Let a be a cardinal. We say that a is singular if it can be represented as

$$a = \sum_{i \in I} a_i$$

where $card(I)$ and all cardinals a_i $(i \in I)$ are strictly less than a.

A cardinal a is regular if it is not singular.

Example 1. Among finite cardinal numbers only 0 and 1 are regular. The first infinite cardinal $\omega_0 = card(\mathbf{N})$ is regular, too. The same is true for the first uncountable cardinal ω_1. More generally, any cardinal of the form $\omega_{\alpha+1}$ is regular. On the other hand, it can easily be observed that any cardinal of the form $\omega_{\alpha+\omega}$ is singular.

Definition 3. An infinite cardinal ω_α is inaccessible if it is regular and its ordinal index α does not have a predecessor, i.e. there is no ordinal β such that $\alpha = \beta + 1$.

For example, in accordance with Definition 3, the first infinite cardinal ω_0 is inaccessible. Actually, this is a unique example (in the theory **ZFC**) of an inaccessible cardinal, because, as follows from one old result of Gödel, there are some models of set theory in which there is no inaccessible cardinal (see, e.g., [99] or [130]).

An infinite cardinal a is called dominant if

$$(\forall b)(b < a \Rightarrow 2^b < a).$$

Definition 4. An infinite cardinal a is strongly inaccessible if it is inaccessible and dominant.

Obviously, ω_0 is strongly inaccessible. As mentioned above, this is a unique example of a strongly inaccessible cardinal in **ZFC**.

Example 2. Let a be an arbitrary infinite cardinal. We define by recursion:

$$a_0 = a, \quad a_{n+1} = 2^{a_n} \quad (n \in \mathbf{N})$$

and, finally, put

$$a^* = \sum_{n \in \mathbf{N}} a_n.$$

It can easily be checked that the cardinal a^* is dominant. Thus, we see that, for any given cardinal a, there exists a dominant cardinal strictly greater than a.

The Generalized Continuum Hypothesis formulated as

$$(\forall \alpha)(2^{\omega_\alpha} = \omega_{\alpha+1})$$

enables to establish the identity between inaccessible and strongly inaccessible cardinal numbers. Namely, it is not hard to prove the following statement.

Theorem 1. *If the Generalized Continuum Hypothesis holds, then these two assertions are equivalent:*

1) a cardinal a is inaccessible;

2) a cardinal a is strongly inaccessible.

We leave the proof of Theorem 1 to the reader, as an exercise.

Now, let us present some typical results concerning the notions of real-valued measurability and two-valued measurability of infinite cardinals.

Theorem 2. *Suppose that the cardinality continuum is not real-valued measurable and let a be an arbitrary infinite cardinal. Then the following two assertions are equivalent:*

1) a is two-valued measurable;

2) a is real-valued measurable.

Theorem 3. *If a cardinal ω_α is not real-valued measurable, then the next cardinal $\omega_{\alpha+1}$ is not real-valued measurable.*

If a cardinal ω_α is not two-valued measurable, then the cardinal 2^{ω_α} is not two-valued measurable.

Theorem 4. *Let $(a_i)_{i\in I}$ be a family of cardinals, satisfying the conditions:*

1) card(I) is not two-valued (real-valued) measurable cardinal;

2) each cardinal a_i ($i \in I$) is not two-valued (real-valued) measurable.

Then the sum $\sum_{i\in I} a_i$ is not two-valued (real-valued) measurable.

Theorems 2 - 4 easily imply the next two statements:

(a) any cardinal number strictly less than the first uncountable inaccessible cardinal is not real-valued measurable;

(b) any cardinal number strictly less than the first uncountable strongly inaccessible cardinal is not two-valued measurable.

All the results presented above are due to Ulam (see his classical paper [194]). Actually, this paper began the cycle of works devoted to large cardinals and other related topics of combinatorial set theory (for more information, see e.g. [78], [89], [99] and [130]). In particular, some further generalizations of statements (a) and (b) can be found in the mentioned works.

Example 3. The cardinality continuum $\mathbf{c} = 2^{\omega_0}$ is not two-valued measurable (according to Theorem 3). The first uncountable cardinal ω_1 is not real-valued measurable (according to the same theorem).

Let us consider the following set-theoretical assertion:

"all infinite cardinals are not real-valued measurable".

Statement (a) immediately implies that this assertion does not contradict the usual axioms of set theory. On the other hand, it is not known at the present time whether the following assertion contradicts the axioms of **ZFC**:

"there exists a real-valued measurable cardinal".

Most specialists in mathematical logic and axiomatic set theory believe that the latter assertion is consistent with **ZFC**. Moreover, consider the following stronger assertion:

"there exists a two-valued measurable cardinal".

Experts working in foundations of mathematics suppose that even the assertion above is consistent with **ZFC**. In this direction, Solovay showed that the next three statements are equiconsistent with **ZFC**:

1) $\mathbf{c}$ is real-valued measurable;
2) there exists a real-valued measurable cardinal;
3) there exists a two-valued measurable cardinal.

In other words, the consistency (with **ZFC**) of each from these three statements implies the consistency of the other ones. For more details, see the indicated works.

Let us also point out that, as follows from the Ulam results formulated above, the first real-valued (two-valued) measurable cardinal is necessarily inaccessible (strongly inaccessible). Later on, it was also demonstrated that there are many strongly inaccessible cardinals strictly less than the first two-valued measurable cardinal (see, e.g., [99], [130] and [132]).

It must be noted that the existence of measurable cardinals (and of other so-called "large" cardinals) yields some nontrivial facts for the structure of the real line (cf. [89], [99]). Of course, the main unsolved problem here is the question whether $\mathbf{c}$ is real-valued measurable. This old problem was first investigated in the classical paper by Banach and Kuratowski [11]. By using some combinatorial argument they demonstrated that, under the Continuum Hypothesis, $\mathbf{c}$ is not real-valued measurable. Obviously, their result trivially follows from the Ulam result stating that ω_1 is not real-valued measurable (let us emphasize once more that the latter result is established within the theory **ZFC**). As mentioned earlier, we are not able to disprove the real-valued measurability of $\mathbf{c}$ within **ZFC**. Moreover, we know that this problem is as difficult as the problem of disproving the existence of a two-valued measurable cardinal.

Here we wish to discuss the question of the real-valued measurability of $\mathbf{c}$ from the topological point of view. Namely, we shall show deep connections of this question with some uncountable subsets of the real line $\mathbf{R}$, having rather strange topological structure.

All topological spaces E considered below are assumed to satisfy the following property: any one-element subset of E is Borel in E. In particular, the spaces in which all one-element subsets are closed possess this property.

Definition 5. We say that a topological space E is a Luzin space if each first category subset of E is at most countable.

It is not hard to prove the following statement.

Theorem 5. *Let E be an arbitrary separable metrizable Luzin space and let μ be a σ-finite diffused Borel measure on E. Then μ is identically zero.*

Proof. Indeed, it is well known that such a measure is concentrated on some first category Borel subset X of E (cf. Exercise 1 from Section 8). Since E is a Luzin space, X must be at most countable. Consequently, taking into account the diffusedness of μ, we have $\mu(X) = 0$. Thus, we may write

$$\mu(E) = \mu(X) + \mu(E \setminus X) = 0 + 0 = 0,$$

and the theorem is proved.

Theorem 5 can also be formulated in the following manner: any separable metrizable Luzin space is universally negligible (from the measure-theoretical point of view).

The next statement plays the key role in our further considerations and is due to Luzin and Mahlo (see, e.g., [141]).

Theorem 6. *Suppose that the Continuum Hypothesis holds. Then there exists a subset E of $\mathbf{R}$ satisfying the relations:*

1) $card(E) = \mathbf{c}$;

2) E is everywhere dense in $\mathbf{R}$;

3) for any first category set $X \subset \mathbf{R}$, the set $X \cap E$ is at most countable.

Proof. Since, according to our assumption, the equality $\mathbf{c} = \omega_1$ is true, we can represent the family of all closed nowhere dense subsets of $\mathbf{R}$ in the form of an ω_1-sequence $(F_\xi)_{\xi<\omega_1}$. Now, let us define by the method of transfinite recursion an injective family $(e_\xi)_{\xi<\omega_1}$ of points of $\mathbf{R}$. Suppose that, for an ordinal $\xi < \omega_1$, the partial family $(e_\zeta)_{\zeta<\xi}$ has already been defined. Consider the set

$$D_\xi = \cup\{F_\zeta \ : \ \zeta < \xi\} \cup \{e_\zeta \ : \ \zeta < \xi\}.$$

Obviously, D_ξ is of first category in $\mathbf{R}$. Hence, in view of the classical Baire theorem,

$$\mathbf{R} \setminus D_\xi \neq \emptyset.$$

Let e_ξ be any point from $\mathbf{R} \setminus D_\xi$. Proceeding in this manner, we are able to construct the entire family $(e_\xi)_{\xi<\omega_1}$. Finally, we put

$$E = \mathbf{Q} \cup \{e_\xi \ : \ \xi < \omega_1\}$$

where $\mathbf{Q}$ denotes, as usual, the set of all rational numbers. Now, it can easily be observed that E is the required subset of $\mathbf{R}$. Indeed, relations 1) and 2) are trivially satisfied. Relation 3) follows directly from the construction of E. Theorem 6 has thus been proved.

Theorem 7. *Let E be a subset of $\mathbf{R}$ described in Theorem 6. Equip E with the topology induced by the standard topology of $\mathbf{R}$. Then E becomes a Luzin space.*

Proof. It suffices to demonstrate that each closed nowhere dense subset of E is at most countable. Let Z be any closed nowhere dense set in E. In virtue of the definition of the induced topology, there exists a closed set $X \subset \mathbf{R}$ for which we have $X \cap E = Z$. Let us check that X is nowhere dense in $\mathbf{R}$. Suppose otherwise, i.e. suppose that there exists a nonempty open subinterval J of $\mathbf{R}$ satisfying the inclusion

$$J \subset int(X).$$

Then, taking into account the fact that E is dense in $\mathbf{R}$, we get

$$J \cap E \subset Z, \quad J \cap E \neq \emptyset,$$

which contradicts the assumption that Z is nowhere dense in E. The contradiction obtained shows us that the set X is necessarily nowhere dense in $\mathbf{R}$. Then, in accordance with relation 3) of Theorem 6, we must have

$$card(Z) = card(X \cap E) \leq \omega,$$

which completes the proof.

Thus, we conclude that if the Continuum Hypothesis holds, then there exists a Luzin subspace of $\mathbf{R}$ having the cardinality continuum. As pointed out above, any σ-finite diffused Borel measure on such a subspace is identically equal to zero. Note that, for the validity of this result, the whole power of the Continuum Hypothesis is not necessary. More precisely, various set-theoretical assertions are known, which are essentially weaker than the Continuum Hypothesis but enable to obtain a result similar to the previous one. In particular, Martin's Axiom (see, e.g., [89], [99], [130]) implies the existence of so-called generalized Luzin subsets of $\mathbf{R}$ for which we have some analogue of the above-mentioned result.

The next statement may be regarded as a purely measure-theoretical consequence of Theorem 7.

Theorem 8. *Under the Continuum Hypothesis, there exists a countably generated σ-algebra S of subsets of $\mathbf{R}$, such that:*

1) all one-element subsets of $\mathbf{R}$ belong to S;

2) there is no nonzero σ-finite diffused measure on $\mathbf{R}$ whose domain contains S.

Proof. Let E be an uncountable Luzin subspace of $\mathbf{R}$ (its existence follows from Theorems 6 and 7). Since we have $card(E) = \mathbf{c}$, there is a bijection

$$f : E \rightarrow \mathbf{R}.$$

Take any countable base $(U_n)_{n \in \mathbf{N}}$ of topology of E and denote by S the σ-algebra of subsets of $\mathbf{R}$, generated by the family $(f(U_n))_{n \in \mathbf{N}}$. It can easily be verified that S is the required σ-algebra.

We see that, under the Continuum Hypothesis, the cardinal $\mathbf{c}$ is not real-valued measurable and, moreover, there exists a countably generated σ-algebra of subsets of $\mathbf{c}$, which contains all one-element subsets of $\mathbf{c}$ but admits no nonzero σ-finite diffused measure. In this connection, let us remark that the result remains true under Martin's Axiom (which is weaker than the Continuum Hypothesis).

Luzin subspaces of $\mathbf{R}$ possess also many other interesting properties. We shall briefly discuss only two of such properties. First, we wish to recall the following important notion from topological measure theory.

Definition 6. Let X be a topological space and let Y be a subset of X. We say that Y is universally measurable in X if, for any Borel probability measure μ given on X, we have $Y \in dom(\mu')$ where μ' denotes the completion of μ.

Example 4. Let X be an arbitrary Polish topological space and let Y be an analytic subset of X. Then, according to the well-known Choquet theorem on capacities (see, e.g., [51], [55] or [155]), Y is universally measurable in X. Actually, this result was established by Luzin and Sierpiński many years ago.

Observe that the family of all universally measurable subsets of a given topological space X forms a σ-algebra in X. This σ-algebra is usually called the σ-algebra of universally measurable sets in X.

In view of Example 4, the σ-algebra generated by all analytic subsets of a Polish space X is contained in the σ-algebra of universally measurable subsets of X. This fact is not so surprising because analytic sets have rather good descriptive structure (see, for instance, [106]). It turns out that any Luzin subset of the real line is universally measurable. In other words, the following statement is true.

Theorem 9. *Let E be a subset of* $\mathbf{R}$ *satisfying the relations of Theorem 6. Then E is universally measurable in* $\mathbf{R}$.

Proof. Take any Borel probability measure μ on $\mathbf{R}$. Without loss of generality, we may assume that μ is diffused (because the family of all those one-element subsets of $\mathbf{R}$ on which μ does not vanish is at most countable). Then, obviously, we have $\mu^*(E) = 0$ (otherwise, there exists a nonzero σ-finite diffused Borel measure on E, which is impossible in view of Theorem 5). Thus, E belongs to the domain of the completion of μ, i.e. E is universally measurable. This ends the proof of the statement.

In particular, we infer that the σ-algebra generated by the family of all Luzin subsets of $\mathbf{R}$ is contained in the σ-algebra of universally measurable sets in $\mathbf{R}$. This shows that Luzin sets are rather good from the point of view of topological measure theory. On the other hand, they turn out to be extremely bad from the point of view of the so-called Baire property. Let us recall the corresponding definition.

Definition 7. Let X be a topological space and let Y be a subset of X. We say that Y has the Baire property (in X) if Y can be represented in the form

$$Y = (U \cup D') \setminus D''$$

where U is an open set in X and D' and D'' are some first category subsets of X.

The Baire property may be regarded as a topological analogue of measurability. This property is thoroughly discussed in [131] and [159]. Note especially that in the excellent textbook by Oxtoby [159] numerous facts and statements for measures and the Baire property are presented and compared. Let us also remark that the existence of subsets of $\mathbf{R}$ which do not have the Baire property (as well as the existence of subsets of $\mathbf{R}$ nonmeasurable in the Lebesgue sense) cannot be established without the aid of uncountable forms of the Axiom of Choice. This follows from the famous result of Solovay (see, e.g., [99]).

Let X be a topological space. We say that a function

$$f : X \to \mathbf{R}$$

has the Baire property if, for any open set $V \subset \mathbf{R}$, the preimage $f^{-1}(V)$ has the Baire property in X.

This definition describes, in purely topological terms, a certain analogue of measurable real-valued functions. Note that functions utilized in real analysis are, as a rule, measurable in the Lebesgue sense and possess the Baire property. Notice also that, for functions possessing the Baire property, a result similar to the classical Luzin theorem (concerning the structure of Lebesgue measurable functions) holds true. More precisely, let X be a topological space and let

$$g : X \to \mathbf{R}$$

be a function. Then the following two assertions are equivalent:

(i) g has the Baire property;

(ii) there exists a first category subset Y of X such that the restriction of g to the set $X \setminus Y$ is continuous.

Obviously, assertion (ii) may be regarded as some analogue of the classical (C)-property of Lebesgue measurable functions (this property is also valid for all real-valued functions measurable with respect to a fixed Radon probability measure on X).

However, there are situations where the analogy between measure and category fails to be true, and the reader must be careful dealing with these two concepts (cf. the corresponding counterexamples considered in the above-mentioned book [159]).

It can easily be checked that the family of all subsets of a topological space X, possessing the Baire property, forms a σ-algebra in X. Evidently, this σ-algebra contains the Borel σ-algebra of X.

Example 5. Let X be an arbitrary Polish space and let Y be an analytic subset of X. Then, according to the classical result of Luzin and

Sierpiński, Y has the Baire property in X. Moreover, Y has the Baire property in the restricted sense (for the definition, see, e.g., [131]). It follows from this fact that if X is uncountable, then the σ-algebra of all subsets of X possessing the Baire property is significantly larger than the Borel σ-algebra of X (the same result can also be obtained by comparing the cardinalities of these two σ-algebras).

Example 6. Assume that the Continuum Hypothesis holds true and consider a Luzin subset E of the real line, constructed in the proof of Theorem 6. It can easily be seen that any subset of the space E, having the Baire property in E, is simultaneously Borel in E. Hence the σ-algebra $B(E)$ coincides with the σ-algebra of all subsets of E possessing the Baire property. Conversely, assuming again the Continuum Hypothesis, let F be an uncountable everywhere dense subset of $\mathbf{R}$ such that $B(F)$ is identical with the σ-algebra of all subsets of F possessing the Baire property (in F). Then it is not hard to demonstrate that F is a Luzin subspace of $\mathbf{R}$. The proof of this fact is left to the reader.

Theorem 10. *Let E be a subset of $\mathbf{R}$ described in Theorem 6. Then E does not have the Baire property in $\mathbf{R}$.*

Proof. Suppose to the contrary that E has the Baire property in $\mathbf{R}$, i.e. E can be represented in the form

$$E = (U \cup D') \setminus D''$$

where U is an open subset of $\mathbf{R}$ and D' and D'' are some first category subsets of $\mathbf{R}$. Only two cases are possible.

1. $U = \emptyset$. In this case, we have the inclusion $E \subset D'$ and hence E turns out to be an uncountable first category set in $\mathbf{R}$. But this is impossible because, according to relation 3) of Theorem 6, the intersection of E with any first category set in $\mathbf{R}$ is at most countable.

2. $U \neq \emptyset$. In this case, we have the inclusion

$$U \setminus D'' \subset E.$$

The set $U \setminus D''$ contains an uncountable G_δ-subset of $\mathbf{R}$ and, consequently, contains a homeomorphic image of the Cantor discontinuum, which is an uncountable nowhere dense closed subset of $\mathbf{R}$. We obtain again that the intersection of E with some first category subset of $\mathbf{R}$ is uncountable, which is impossible.

Thus, in the both cases, we get a contradiction which shows that E cannot possess the Baire property in $\mathbf{R}$. Actually, the same argument applied to any uncountable subset Y of E leads to the conclusion that Y does not possess the Baire property in $\mathbf{R}$.

The preceding theorem also yields the following result: under the Continuum Hypothesis, there are universally measurable subsets of $\mathbf{R}$ which do not have the Baire property.

On the other hand, it is not hard to prove that there exist first category subsets of $\mathbf{R}$ (hence having the Baire property) which are not measurable in the Lebesgue sense. Indeed, as we know, the standard Lebesgue measure λ_1 is concentrated on some first category subset Z of $\mathbf{R}$. In particular, $\lambda_1(Z) > 0$. This immediately implies (cf. [154] or [159]) that Z contains a λ_1-nonmeasurable subset which simultaneously is of first category (consequently, has the Baire property). Analogously, it can be proved, by applying a dual argument, that there are subsets of $\mathbf{R}$ of Lebesgue measure zero which do not possess the Baire property.

The following question arises naturally: does there exist a nonzero σ-finite diffused measure defined on the σ-algebra of all subsets of $\mathbf{R}$ having the Baire property? This question remains unsolved since it is logically equivalent to the problem of the real-valued measurability of $\mathbf{c}$ (the reader can check the above-mentioned equivalence without any difficulty). At the same time, it was proved that there exists no nonzero σ-finite measure on the σ-algebra of all subsets of $\mathbf{R}$ possessing the Baire property, quasi-invariant with respect to an uncountable group of translations of $\mathbf{R}$ (for more details, see [117]). Note that the proof of this result is essentially based on the Ulam theorem stating that the first uncountable cardinal ω_1 is not real-valued measurable (see Theorem 3).

In connection with the non-real-valued measurability of ω_1, one important result of Luzin must be noted. Namely, by applying some delicate argument from the theory of analytic sets, Luzin was able to prove, in **ZFC**, that there exists a universally negligible subset of $\mathbf{R}$ of cardinality ω_1 (the proof is presented, e.g., in [131] and [149]). This interesting result turns out to be rather sharp because, as shown by Baumgartner and Laver, there are some models of set theory in which every universally negligible subset of $\mathbf{R}$ is of cardinality less than or equal to ω_1 (for more details, see [149] and the references therein).

We want to conclude this Appendix with some remarks illustrating the role of measurable cardinals in general topology (more precisely, in those questions of topology which arise when various generalizations of the fundamental concept of quasicompactness are examined).

Definition 8. Let a be an infinite cardinal. We say that a topological space X is a-quasicompact if any open covering of X contains a subcovering whose cardinality is strictly less than a.

In particular, if $a = \omega_0$, we come to the standard notion of a quasicompact space (we recall that a topological space is compact if it is quasicompact and Hausdorff simultaneously).

The classical theorem of Tychonoff (see [193]) states that the topological product of an arbitrary family of quasicompact spaces is quasicompact, too. This result may be regarded as one of the main statements in modern set-theoretical topology (actually, as shown by Kelley, the Tychonoff theorem is logically equivalent to the Axiom of Choice). Thus, the Tychonoff theorem says that, for $a = \omega_0$, the a-quasicompactness is preserved under all products. In this connection, the following result is of some interest (see, e.g., [132]).

Theorem 11. *Let a be an uncountable non-two-valued measurable cardinal. Then there exists a family $(X_i)_{i \in I}$ of a-quasicompact spaces, whose topological product $\prod_{i \in I} X_i$ is not an a-quasicompact space.*

In other words, Theorem 11 says that, only for very large cardinals a, we may hope to obtain an appropriate analogue of the classical Tychonoff theorem in terms of the notion of a-quasicompactness. Since the existence of such cardinals a cannot be established within the theory **ZFC**, any analogue of the Tychonoff theorem for a-quasicompact spaces will be of less importance.

There are also many other interesting examples in various fields of mathematics (primarily, in universal algebra, topology, functional analysis and real analysis), showing the role of measurable cardinals in concrete problems from those fields. We suppose that the material presented in this Appendix is completely sufficient for illustration and vividly demonstrates the significance of the theory of large cardinals in modern mathematics.

Bibliography

[1] Adler R.J., *An Introduction to Continuity and Related Topics for General Gaussian Processes*, Institute of Math. Statistics, Hayward, 1990.

[2] Alexandrov A.D., *To the theory of mixed volumes of convex bodies I*, Mat. Sborn., 2 (44), 1937, pp. 947 - 972 (in Russian).

[3] Alexandrov A.D., *To the theory of mixed volumes of convex bodies II*, Mat. Sborn., 2 (44), 1937, pp. 1205 - 1238 (in Russian).

[4] Alexandrov A.D., *To the theory of mixed volumes of convex bodies III*, Mat. Sborn., 3 (45), 1938, pp. 27 - 46 (in Russian).

[5] Alexandrov A.D., *To the theory of mixed volumes of convex bodies IV*, Mat. Sborn., 3 (45), 1938, pp. 227 - 252 (in Russian).

[6] Alexandrov A.D., *The existence almost everywhere of a second differential of a convex function and some related properties of convex surfaces*, Uch. Zap. LGU, vol. 37, no. 6, 1939, pp. 3 - 35 (in Russian).

[7] Alexandrov A.D., *Convex Polyhedra*, Gostechizdat, Moscow, 1950 (in Russian).

[8] Anderson T., *The integral of symmetric unimodal functions over a symmetric convex set and some probability inequalities*, Proc. Amer. Math. Soc., vol. 6, no. 2, 1955, pp. 170 - 176.

[9] Baker R., *"Lebesgue measure" on* $\mathbf{R}^\infty$, Proc. Amer. Math. Soc., vol. 113, no. 4, 1991, pp. 1023 - 1029.

[10] Banach S., *Théorie des Opérations Linéaires*, Monogr. Matem., Warsaw, 1932.

[11] Banach S., Kuratowski C., *Sur une généralisation du probléme de la mesure*, Fund. Math., vol. 14, 1929, pp. 127 - 131.

[12] Banach S., Tarski A., *Sur la décomposition des ensembles de points en parties respectivement congruentes*, Fund. Math., vol. 6, 1924, pp. 244 - 277.

[13] Belyaev Yu.K., *Continuity and Hölder's conditions for sample functions of stationary Gaussian processes*, Proc. 4-th Berkeley Symp. on Mathematical Statistics and Probability, vol. 2, 1961, pp. 23 - 33.

[14] Billingsley P., *Convergence of Probability Measures*, Wiley and Sons, New York, 1968.

[15] Birkhoff G., *Tres observaciones sobre el algebra lineal*, Univ. Nac. Tucumán Rev., ser. A, vol. 5, 1946, pp. 147 - 151.

[16] Blaschke W., *Vorlesungen über Differentialgeometrie*, volumes 1 - 3, Berlin, 1921, 1923, 1929.

[17] Blaschke W., *Vorlesungen über Integralgeometrie*, Berlin, 1955.

[18] Blaschke W., *Kreis und Kugel*, W. de Gruyter, Berlin, 1956.

[19] Bogachev V.I., *Locally convex spaces with CLT property and supports of measures*, Vestn. MGU, Math. Mech. Ser., no. 6, 1986, pp. 16 - 20 (in Russian).

[20] Bogachev V.I., *Gaussian Measures*, Izd. Nauka, Moscow, 1997 (in Russian).

[21] Bonnesen T., Fenchel W., *Theorie der Konvexen Körper*, Springer Verlag, Berlin, 1934.

[22] Borell C., *Convex measures on locally convex spaces*, Ark. Mat., vol. 12, no. 2, 1974, pp. 239 - 252.

[23] Borell C., *The Brunn-Minkowski inequality in Gauss space*, Invent. Math., vol. 30, no. 2, 1975, pp. 207 - 216.

[24] Borell C., *Gaussian Radon measures on locally convex spaces*, Math. Scand., vol. 38, no. 2, 1976, pp. 265 - 284.

[25] Borell C., *Capacitary inequalities of the Brunn-Minkowski type*, Math. Ann., vol. 263, no. 2, 1983, pp. 179 - 184.

[26] Borevich Z.I., Shafarevich I.R., *Number Theory*, Izd. Nauka, Moscow, 1972 (in Russian).

[27] Borovikov V., *On the intersection of a sequence of simplices*, Usp. Mat. Nauk, vol. 7, no. 6 (52), 1952, pp. 179 - 180 (in Russian).

[28] Borovkov A.A., *Convergence of measures and stochastic processes*, Usp. Mat. Nauk, vol. 31, no. 2, 1976, pp. 3 - 68 (in Russian).

[29] Bourbaki N., *Topological Vector Spaces*, Izd. Nauka, Moscow, 1959 (in Russian, translation from French).

[30] Bourbaki N., *Set Theory*, Izd. Mir, Moscow, 1965 (in Russian, translation from French).

[31] Bourbaki N., *Integration*, Izd. Mir, Moscow, 1977 (in Russian, translation from French).

[32] Brascamp H., Lieb E.H., *On extensions of the Brunn-Minkowski and Precopa-Leindler theorems, including inequalities for logconcave functions, and with an application to the diffusion equation*, Journ. Funct. Anal., vol. 22, 1976, pp. 366 - 389.

[33] Brunn H., *Über Ovale und Eiflächen*, Inaugural-Diss., München, 1887.

[34] Buldygin V.V., *On the convergence of random sequences with values in a Banach space*, Dokl. Akad. Nauk SSSR, vol. 213, no. 5, 1973, pp. 1012 - 1014 (in Russian).

[35] Buldygin V.V., *On random series in a Banach space*, Teor. Ver. i Prim., vol. 18, no. 3, 1973, pp. 491 - 504 (in Russian).

[36] Buldygin V.V., *On invariant Bayesian estimators for generalized random variables*, Teor. Ver. i Prim., vol. 22, no. 1, 1977, pp. 175 - 179 (in Russian).

[37] Buldygin V.V., *Convergence of Random Elements in Topological Spaces*, Izd. Naukova Dumka, Kiev, 1980 (in Russian).

[38] Buldygin V.V., *Random elements and their convergence in topological spaces*, Teor. Ver. i Prim., vol. 26, no. 1, 1981, pp. 88 - 100 (in Russian).

[39] Buldygin V.V., *On a comparison inequality for the distribution of the maximum of a Gaussian process*, Teor. Ver. i Mat. Statist., no. 28, 1983, pp. 9 - 14 (in Russian).

[40] Buldygin V.V., *Supports of probability measures in separable Banach spaces*, Teor. Ver. i Prim., vol. 29, no. 3, 1984, pp. 528 - 532 (in Russian).

[41] Buldygin V.V., *On the convergence of Fourier series of stationary Gaussian processes*, Ukr. Mat. Journ., vol. 39, no.3, 1987, pp. 278 - 282 (in Russian).

[42] Buldygin V.V., Kharazishvili A.B., *On Borel measures in nonseparable metric spaces*, Ukr. Mat. Journ., vol. 29, no. 4, 1983, pp. 552 - 556 (in Russian).

[43] Buldygin V.V., Kharazishvili A.B., *Anderson inequality and unimodal measures*, Teor. Ver. i Mat. Stat., vol. 35, 1986, pp. 13 - 27 (in Russian).

[44] Buldygin V.V., Kharazishvili A.B., *Brunn-Minkowski Inequality and its Applications*, Izd. Naukova Dumka, Kiev, 1985 (in Russian).

[45] Buldygin V.V. Solntsev S.A., *On the oscillation of realizations of Gaussian sequences bounded almost everywhere*, Ukr. Mat. Journ., vol. 37, no. 1, 1985, pp. 110 - 111 (in Russian).

[46] Buldygin V., Solntsev S., *Asymptotic Behaviour of Linearly Transformed Sums of Random Variables*, Kluwer Academic Publishers, Dordrecht, 1997.

[47] Burago D.M., Zalgaller V.A., *Geometric Inequalities*, Izd. Nauka, Leningrad, 1980 (in Russian).

[48] Busemann H., *Convex Surfaces*, Interscience, New York, 1958.

[49] Cassels J.W.S., *An Introduction to the Geometry of Numbers*, Springer Verlag, Berlin, 1959.

[50] Chandrasekharan K., *Introduction to Analytic Number Theory*, Springer Verlag, Berlin, 1968.

[51] Choquet G., *Theory of capacities*, Ann. Inst. Fourier, vol. 5, 1953 - 1954, pp. 131 - 295.

[52] Cichoń J., Kharazishvili A., Weglorz B., *Subsets of the Real Line*, Lódź University Press, Lódź, 1995.

[53] Daletski I.L., Fomin S.V., *Measures and Differential Equations in Infinite-dimensional Spaces*, Izd. Nauka, Moscow, 1983 (in Russian).

[54] Davis W.J., Figiel T., Johnson W.B., Pelczyński A., *Factoring weakly compact operators*, Journ. Func. Anal., vol. 17, 1974, pp. 311 - 323.

[55] Dellacherie C., *Capacités et Processus Stochastiques*, Springer Verlag, Berlin, 1972.

[56] Diestel J., *Geometry of Banach Spaces*, Springer Verlag, Berlin, 1975.

[57] Doob J.L., *Stochastic Processes*, Wiley and Sons, New York, 1953.

[58] Driml M., Hansh O., *Conditional expectation for generalized random variables*, Trans. 2-nd Prague Conference on Information Theory, Statistical Decision Functions and Random Processes, Prague, 1960, pp. 123 - 144.

[59] Dunford N., Schwartz J.T., *Linear Operators*, Part 1, Interscience Publishers, New York, 1958.

[60] Dye H.A., *On groups of measure preserving transformations. I*, Amer. Journ. Math., vol. 81, no. 1, 1959, pp. 119 - 159.

[61] Dye H.A., *On groups of measure preserving transformations. II*, Amer. Journ. Math., vol. 85, no. 4, 1963, pp 551 - 576.

[62] Eaton M.L., *A review of selected topics in multivariate probability inequalities*, Ann. Statist., vol. 10, 1982, pp. 11 - 43.

[63] Eggleston H., *Convexity*, Cambridge University Press, Cambridge, 1958.

[64] Egorychev G.P., *A proof of the van der Waerden conjecture for permanents*, Sib. Mat. Journ., vol. 22, no. 6, 1981, pp. 65 - 71 (in Russian).

[65] Ehrhard A., *Symetrisation dans l'espace de Gauss*, Math. Scand., vol. 53, 1983, pp. 281 - 301.

[66] Ehrhard A., *Un principe de symetrisation dans les espaces de Gauss*, Lect. Notes in Math., v. 990, 1983, pp. 92 - 101.

[67] Ehrhard A., *Elements extremaux pour les inegalites de Brunn-Minkowski gaussiennes*, Ann. Inst. H. Poincare, vol. 22, no. 1, 1986, pp. 149 - 168.

[68] Erdös P., *On the law of the iterated logarithm*, Ann. of Math., vol. 43, 1942, pp. 419 - 436.

[69] Ershov M.P., *Measure extensions and stochastic equations*, Teor. Ver. i Prim., vol. 19, no. 3, 1974 (in Russian).

[70] Falikman D.I., *A proof of the van der Waerden conjecture on permanents of doubly stochastic matrices*, Mat. Zam., vol. 29, no. 6, 1981, pp. 931 - 938 (in Russian).

[71] Federer H., *Geometric Measure Theory*, Springer Verlag, Berlin, 1969.

[72] Fejes Tóth L., *Lagerungen in der Ebene auf der Kugel und im Raum*, Springer Verlag, Berlin, 1953.

[73] Feller W., *The general form of the so-called law of the iterated logarithm*, Trans. Amer. Math. Soc., vol. 54, 1943, pp. 373 - 402.

[74] Feller W., *An Introduction to Probability Theory and its Applications*, volumes I - II, Wiley and Sons, New York, 1970 - 1971.

[75] Fernique X., *The regularity of trajectories of Gaussian random functions*, in: Stochastic Processes, vol. 10, Izd. Mir, Moscow, 1978, pp. 63 - 132 (in Russian, translation from French).

[76] Figiel T., *Factorization of compact operators and applications to the approximation problems*, Studia Math., vol. 45, 1973, pp. 191 - 210.

[77] Finkelstein H., *The law of the iterated logarithm for empirical distributions*, Ann. Math. Statist., vol. 42, no.2, 1971, pp. 607 - 615.

[78] Fremlin D.H., *Real-valued-measurable cardinals*, Israel Math. Conf. Proc., vol. 6, 1993, pp. 151 - 304.

[79] Fremlin D.H., Talagrand M., *A Gaussian measure on l_∞*, Ann. Probab., vol. 8, no. 6, 1980, pp. 1192 - 1193.

[80] Gikhman J.J., Skorokhod A.V., *The Theory of Stochastic Processes*, vol. 1, Izd. Nauka, Moscow, 1971 (in Russian).

[81] Gnedenko B.V., Kolmogorov A.N., *Limit Distributions for Sums of Independent Random Variables*, Gostechizdat, M. - L., 1949 (in Russian).

[82] Gordon Y., *Some inequalities for Gaussian processes and applications*, Israel Journ. Math., vol. 50, 1985, pp. 265 - 289.

[83] Gruber P.M., *Über den Durchschnitt einer abnehmenden Folge von Parallelepipeden*, Elemente der Mathematik, vol. 32, no. 1, 1977, pp. 13 - 15.

[84] Grünbaum B., *Convex Polytopes*, Wiley and Sons, New York, 1967.

[85] Hadwiger H., *Vorlesungen über Inhalt, Oberfläche und Isoperimetrie*, Springer Verlag, Berlin, 1957.

[86] Halmos P.R., *Measure Theory*, D.Van Nostrand, New York, 1950.

[87] Halmos P.R., *Lectures on Ergodic Theory*, The Mathematical Society of Japan, Tokyo, 1956.

[88] *Handbook of Convex Geometry*, (ed. by Gruber P.M.), volumes 1 - 2, North-Holland Publ. Comp., Amsterdam, 1990 - 1991.

[89] *Handbook of Mathematical Logic*, (ed. by Barwise J.), North-Holland Publ. Comp., Amsterdam, 1977.

[90] Hausdorff F., *Bemerkung über den Inhalt von Punktmengen*, Math. Ann., vol. 75, 1914, pp. 428 - 433.

[91] Henstock R., Macbeath A.M., On the measure of sum-sets I: The theorems of Brunn, Minkowski and Lusternik, Proc. London Math. Soc., III Ser., vol. 3, 1953, pp. 182 - 194.

[92] Hopf E., *Theory of measures and invariant integrals*, Trans. Amer. Math. Soc., vol. 34, 1932, pp. 373 - 393.

[93] Hopf E., *Ergodentheorie*, Springer Verlag, Berlin, 1937.

[94] Huffer F.W., *Slepian's inequality via the central limit theorem*, Canad. Journ. of Statistics, vol. 14, no. 4, 1986, pp. 367 - 370.

[95] Iaglom I.M., Boltyanskii V.G., *Convex Figures*, Gostechizdat, M. - L., 1951 (in Russian).

[96] Ibragimov I.A., Haśminskii R.Z., *Asymptotic Theory of Estimation*, Izd. Nauka, Moscow, 1979 (in Russian).

[97] Ito K., Nisio M., *On the oscillation functions of Gaussian processes*, Math. Scand., vol. 22, no. 1, 1968, pp. 209 - 223.

[98] Jain N., Kallianpur G., *Oscillation function of a multiparameter Gaussian process*, Nagoya Math. Journ., vol. 47, 1972, pp. 15 - 28.

[99] Jech T., *Set Theory*, Academic Press, New York and London, 1978.

[100] John K., Zizler V., *A renorming of dual spaces*, Israel Journ. Math., vol. 12, 1972, pp. 331 - 336.

[101] John K., Zizler V., *A note on renorming of dual spaces*, Bull. Acad. Polon. Sci., vol. 21, 1973, pp. 47 - 50.

[102] Johnson W.B., *Factoring compact operators*, Israel Journ. Math., vol. 9, 1971, pp. 337 - 345.

[103] Kahane J.-P., *Some Random Series of Functions*, D.C. Heath and Comp., Massachusetts, 1968.

[104] Kahane J.-P., *Une inegalite du type de Slepian et Gordon sur les processus gaussiennes*, Israel Journ. Math., vol. 55, 1986, pp. 109 - 110.

[105] Kakutani S., *Two fixed-point theorems concerning bicompact convex sets*, Proc. Imp. Acad. Tokyo, vol. 14, 1938, pp. 242 - 245.

[106] Kechris A.S., *Classical Descriptive Set Theory*, Springer Verlag, New York, 1995.

[107] Kelley J.L., *General Topology*, D.Van Nostrand, New York, 1955.

[108] Kesten H., *The limit points of a normalized random walk*, Ann. Math. Statist., vol. 41, no. 4, 1970, pp. 1173 - 1205.

[109] Kharazishvili A.B., *Characteristic properties of parallelepipeds*, Soob. Akad. Nauk Gruz. SSR, vol. 72, no. 1, 1973, pp. 17 - 19 (in Russian).

[110] Kharazishvili A.B., *On locally conic sets*, Soob. Akad. Nauk Gruz. SSR, vol. 80, no. 1, 1975, pp. 29 - 32 (in Russian).

[111] Kharazishvili A.B., *Introduction to Combinatorial Geometry*, Izd. Tbil. Gos. Univ., Tbilisi, 1985 (in Russian).

[112] Kharazishvili A.B., *Elements of the Combinatorial Theory of Infinite Sets*, Izd. Tbil. Gos. Univ., Tbilisi, 1981 (in Russian).

[113] Kharazishvili A.B., *Invariant Extensions of Lebesgue Measure*, Izd. Tbil. Gos. Univ., Tbilisi, 1983 (in Russian).

[114] Kharazishvili A.B., *On the existence of quasi-invariant measures*, Soob. Akad. Nauk Gruz. SSR, vol. 115, no. 1, 1984, pp. 37 - 40 (in Russian).

[115] Kharazishvili A.B., *Some problems in measure theory*, Coll. Math., vol. LXII, no. 2, 1991, pp. 197 - 220.

[116] Kharazishvili A.B., *On the existence of estimates for strongly separated families of probability measures*, Trudy Mat. Inst. im. A.M. Razmadze, vol. 92, 1989, pp. 100 - 105 (in Russian).

[117] Kharazishvili A.B., *Transformation Groups and Invariant Measures*, World Scientific Publ. Comp., Singapore, 1998.

[118] Khintchin A., *Über einen Satz der Wahrscheinlichkeitsrechnung*, Fund. Math., vol. 6, 1924, pp. 9 - 20.

[119] Khinchin A., *On unimodal distributions*, Trudy NIIMM Tomsk. Gos. Univ., vol. 2, 1938, pp. 1 - 6 (in Russian).

[120] Klee V.L., *Some characterizations of convex polyhedra*, Acta Math., vol. 102, 1959, pp. 79 - 107.

[121] Kolmogoroff A.N., *Grundberiffe der Wahrscheinlichkeitsrechnung*, Springer Verlag, Berlin, 1933.

[122] Kolmogoroff A.N., *Zur Normierbarkeit eines allgemeinen topologischen linearen Raumes*, Studia Math., vol. 5, 1934, pp. 29 - 33.

[123] Kolmogoroff A.N., *La transformation de Laplace dans les espaces linéaires*, C. R. Acad. Sci. Paris, vol. 200, 1935, pp. 1717 - 1718.

[124] Kolmogorov A.N., Fomin S.V., *Elements of the Theory of Functions and Functional Analysis*, Izd. Nauka, Moscow, 1981 (in Russian).

[125] Kornfeld I.P., Sinai Ya.G., Fomin S.V., *Ergodic Theory*, Izd. Nauka, Moscow, 1980 (in Russian).

[126] Krein M., Milman D., *On extreme points of regularly convex sets*, Studia Math., vol. 9, 1940, pp. 133 - 138.

[127] Krieger W., *On non-singular transformations of a measure space. I*, Z. Wahrscheinlichkeitstheorie und verw. Geb., vol. 11, no. 2, 1969, pp. 83 - 97.

[128] Krieger W., *On non-singular transformations of a measure space. II*, Z. Wahrscheinlichkeitstheorie und verw. Geb., vol. 11, no. 2, 1969, pp. 98 - 119.

[129] Kryloff N., Bogoliouboff N.O., *La theorie generale de la measure dans son application a l'etude des systemes dynamiques de la mecanique non lineaire*, Ann. Math., vol. 38, 1937, pp. 65 - 113.

[130] Kunen K., *Set Theory*, North-Holland Publ. Comp., Amsterdam, 1980.

[131] Kuratowski K., *Topology*, vol. 1, Academic Press, New York and London, 1966.

[132] Kuratowski K., Mostowski A., *Set Theory*, North-Holland Publ. Comp., Amsterdam, 1967.

[133] Kwapień S., Woyczyński W., *Random Series and Stochastic Integrals: Single and Multiple*; Birkhäuser, Basel, 1992.

[134] Ledoux M., *Isoperimetry and Gaussian analysis*, Lect. Notes in Math., vol. 1648, 1996, pp. 165 - 294.

[135] Ledoux M., Talagrand M., *Probability in Banach Spaces*, Springer Verlag, Berlin, 1991.

[136] Lévy P., *Théorie de l'Addition des Variables Aléatoires*, Gauthier-Villars, Paris, 1937.

[137] Lifshits M.A., *Gaussian Random Functions*, Kluwer Academic Publishers, Dordrecht, 1995.

[138] Lindelöf L., *Propriétés générales des polyedres qui, sous une étendue superficielle donnée, renferment le plus grand volume*, Bull. Acad. Sci. St. Pétersbourg, vol. 14, 1869, pp. 257 - 269.

[139] Lomnicki A., *Nouveau fondements du calcul des probabilites*, Fund. Math., vol. 4, 1923, pp. 34 - 71.

[140] Lubotzky A., *Discrete Groups, Expanding Graphs and Invariant Measures*, Birkhäuser, Basel, 1994.

[141] Lusin N., *Sur un probléme de M. Baire*, C.R. Acad. Sci. Paris, vol. 158, 1914, pp. 1258 - 1261.

[142] Lusternik L.A., *The Brunn-Minkowski inequality for arbitrary measurable sets*, Dokl. Akad. Nauk SSSR, vol. 3, 1935, pp. 55 - 58 (in Russian).

[143] Lusternik L.A., *Convex Figures and Polyhedra*, Gostechizdat, Moscow, 1956 (in Russian).

[144] Mackey G.W., *Borel structures in groups and their duals*, Trans. Amer. Math. Soc., vol. 85, 1957, pp. 134 - 169.

[145] Marcus M.B., Shepp L., *Sample behavior of Gaussian processes*, Proc. 6-th Berkeley Symp. Math. Statist. Probab., vol. 2, 1971, pp. 432 - 442.

[146] Marczewski E., Sikorski R., *Measures in nonseparable metric spaces*, Coll. Math., vol. 1, 1948, pp. 133 - 139.

[147] Markov A.A., *Some theorems on Abelian sets*, Doklady Akad. Nauk SSSR, vol. 1, 1936, pp. 299 - 302 (in Russian).

[148] Marshall A.W., Olkin I., *Inequalities: Theory of Majorization and its Applications*, Academic Press, New York, 1979.

[149] Miller A.W., *Special subsets of the real line*, in: Handbook of Set-Theoretic Topology (edited by Kunen K. and Vaughan J.), North-Holland Publ. Comp., Amsterdam, 1984.

[150] Minkowski H., *Allgemeine Lehrsätze über die konvexen Polyeder*, Nachr. Ges. Wiss., Göttingen, Math.-Phys. Kl., 1897, pp. 198 - 219.

[151] Minkowski H., *Geometrie der Zahlen*, Leipzig, 1910.

[152] Minkowski H., *Gesammelte Abhandlungen*, vol. 2, Teubner, Leipzig - Berlin, 1911.

[153] Mushtari D., *Probabilities and Topologies in Banach spaces*, Izd. Kazan. Univ., Kazań, 1989 (in Russian).

[154] Natanson I.P., The Theory of Functions of a Real Variable, Izd. Nauka, Moscow, 1957 (in Russian).

[155] Neveu J., *Bases Mathématiques du Calcul des Probabilités*, Masson et Cie, Paris, 1964.

[156] Ornstein D., Rudolph D., Weiss B., *Equivalence of measure preserving transformations*, Mem. Amer. Math. Soc., vol. 37, no. 262, 1982, pp. 1 - 120.

[157] Ostrovskii E.I., *On supports of probability measures in separable Banach spaces*, DAN SSSR, vol. 255, no. 6, 1980, pp. 836 - 838 (in Russian).

[158] Oxtoby J.C., *Invariant measures in groups which are not locally compact*, Trans. Amer. Math. Soc., vol. 60, 1946, pp. 215 - 237.

[159] Oxtoby J.C., *Measure and Category*, Springer-Verlag, Berlin, 1971.

[160] Oxtoby J.C., Ulam S., *On the existence of a measure invariant under a transformation*, Ann. Math., vol. 40, 1939, pp. 560 - 566.

[161] Parthasarathy K.R., *Probability Measures on Metric Spaces*, Academic Press, New York, 1967.

[162] Petrov V.V., *Sums of Independent Random Variables*, Izd. Nauka, Moscow, 1972 (in Russian).

[163] Phelps R.R., *Lectures on Choquet's Theorem*, Van Nostrand, Princeton, N.J., 1966.

[164] Piterbarg V.I., *Asymptotic Methods in the Theory of Gaussian Processes and Fields*, Izd. MGU, Moscow, 1988 (in Russian).

[165] Polya G., Szegö G., *Isoperimetric Inequalities in Mathematical Physics*, Princeton University Press, Princeton, 1951.

[166] Precopa A., *Logarithmic concave measures with applications*, Acta Sci. Math., vol. 32, 1971, pp. 301 - 316.

[167] Prokhorov Ju.V., *Probability distributions in function spaces*, Usp. Mat. Nauk, vol. 8, 1953, pp. 165 - 166 (in Russian).

[168] Prokhorov Ju.V., *Convergence of random processes and limit theorems in probability theory*, Teor. Ver. i Prim., vol. 1, no. 2, 1956, pp. 177 - 238 (in Russian).

[169] Ramachandran D., *A new proof of Hopf's theorem on invariant measures*, Contemporary Mathematics, vol. 94, 1989, pp. 263 - 270.

[170] Ramachandran D., Misiurewicz M., *Hopf's theorem on invariant measures for a group of transformations*, Studia Math., vol. LXXIV, 1982, pp. 183 - 189.

[171] Rao M.M., *Stochastic Processes and Integration*, Sijthoff and Noordhoff, Alphen aan den Rijn, 1979.

[172] Rasiowa H., Sikorski R., *The Mathematics of Metamathematics*, PAN, Warszawa, 1963.

[173] Rockafellar R.T., *Convex Analysis*, Princeton University Press, Princeton, 1970.

[174] Rogers C.A., Shephard G.C., *The difference body of a convex body*, Arch. Math., vol. 8, 1957, pp. 220 - 233.

[175] Rudin W., *Functional Analysis*, McGraw-Hill Book Comp., New York, 1973.

[176] Sakrison D.J., *Notes on Analog Communication*, D. Van Nostrand, New York, 1970.

[177] Saks S., *Theory of the Integral*, Warszawa-Lwów, 1937.

[178] Santaló L.A., *Integral Geometry and Geometric Probability*, Addison-Wesley Publ. Comp., Inc., Reading, Massachusetts, 1976.

[179] Schaefer H.H., *Topological Vector Spaces*, The Macmillan Comp., New York, 1966.

[180] Shiryaev A.N., *Probability*, Izd. Nauka, Moscow, 1980 (in Russian).

[181] Shoenfield J.R., *Measurable cardinals*, Logic Colloquium'69, North-Holland Publ. Comp., Amsterdam, 1971.

[182] Skorokhod A.V., *Studies in the Theory of Random Processes*, Izd. Kievsk. Gos. Univ., Kiev, 1961 (in Russian).

[183] Skorokhod A.V., *Integration in Hilbert Space*, Izd. Nauka, Moscow, 1975 (in Russian).

[184] Slepian D., *The one-sided barier problem for Gaussian noise*, Bull. Syst. Tech. J., vol. 41, no. 3, 1962, pp. 463 - 501.

[185] Solovay R.M., *Real-valued measurable cardinals*, Proceedings of Symposia in Pure Mathematics, vol. XII, Axiomatic Set Theory, Part 1, Providence 1971, pp. 397 - 428.

[186] Steinhaus H., *Sur les distances des points de mesure positive*, Fund. Math., vol. 1, 1920, pp. 93 - 104.

[187] Strassen V., *An invariance principle for the law of iterated logarithm*, Z. Wahrscheinlichkeitstheorie verw. Gebiete, vol. 3, no. 3, 1964, pp. 211 - 226.

[188] Sudakov V., *Geometric problems of the theory of infinite-dimensional probability distributions*, Trudy Mat. Inst. im. V.A.Steklova, vol. 161, Leningrad, 1976 (in Russian).

[189] Talagrand M., *Mesures gaussiennes sur un espace localement convexe*, Z. Wahrscheinlichkeitstheorie und verw. Geb., vol. 64, 1983, pp. 181 - 209.

[190] Talagrand M., *Regularite des processus gaussiens*, C. R. Acad. Sci. Paris, vol. 301, ser. 1, no. 7, 1985, pp. 379 - 381.

[191] Talagrand M., *La description des processus gaussiens bornes*, C. R. Acad. Sci. Paris, vol. 301, ser. 1, no. 15, 1985, pp. 751 - 753.

[192] Tsirelson B.S., *The natural modification of a stochastic process and its applications to random function series and Gaussian measures*, Zap. Sem. LOMI, vol. 55, 1976, pp. 35 - 63 (in Russian).

[193] Tychonoff A., *Über die topologische Erweiterung von Räumen*, Math. Ann., vol. 102, 1930, pp. 544 - 561.

[194] Ulam S., *Zur Masstheorie in der allgemeinen Mengenlehre*, Fund. Math., vol. 16, 1930, pp. 140 - 150.

[195] Ulam S.M., *A Collection of Mathematical Problems*, Los Alamos Sci. Lab., New Mexico, 1960.

[196] Vakhania N.N., *Probability Distributions in Vector Spaces*, Izd. Metsniereba, Tbilisi, 1971 (in Russian).

[197] Vakhania N.N., Tarieladze V.I., Chobanian S.A., *Probability Distributions in Banach Spaces*, Izd. Nauka, Moscow, 1985 (in Russian).

[198] Vladimirov D.A., *Boolean Algebras*, Izd. Nauka, Moscow, 1969 (in Russian).

[199] Wagon S., *The Banach-Tarski Paradox*, Cambridge University Press, Cambridge, 1985.

[200] Yurinskii V.V., *Sums and Gaussian Vectors*, Springer Verlag, Berlin, 1995.

[201] Zolotarev V.M., *One-dimensional Stable Distributions*, Izd. Nauka, Moscow, 1983 (in Russian).

Subject Index

MIX
Papier aus verantwortungsvollen Quellen
Paper from responsible sources
FSC® C105338

If you have any concerns about our products,
you can contact us on
ProductSafety@springernature.com

In case Publisher is established outside the EU,
the EU authorized representative is:
Springer Nature Customer Service Center GmbH
Europaplatz 3, 69115 Heidelberg, Germany

Printed by Libri Plureos GmbH
in Hamburg, Germany